A Laboratory Text for Developmental Biology

Second Edition

By
Peter B. Armstrong
Carol A. Erickson
Robert D. Grey
Jeaneen K. A. Metzler

University of California, Davis

Layout composition by Martha Spence
Avocet Design

KENDALL/HUNT PUBLISHING COMPANY
4050 Westmark Drive   Dubuque, Iowa 52002

Printed in the United States of America
10 9 8 7 6 5 4 3 2 1

Cover photo: Whole mount chicken embryo, fluorescein labeled with EC-8, an intermediate filament antibody that distinguishes neurites. Photo by Jeanne F. Loring.

An understanding of developmental biology requires not only knowledge about what happens during embryonic development, but also insight into how events occur. The lecture portion of a course in development deals primarily with the latter, presenting theoretical questions of the subject and the experimental approaches to them. In the laboratory the student has the opportunity to become acquainted with the developing organisms and to learn the dynamic and fundamental structural changes that lead to the establishment of the mature form.

In this laboratory text we have attempted to streamline the presentation of embryonic structure by eliminating all but the most basic and fundamental aspects of anatomy. This has been based on the premise that developing embryos, like any biological entity, are not abstractions relegated to textbook diagrams, but are instead real things, literally flesh and blood, and the best way to know how one is put together is to look at it. Reports from the nearly five thousand students who have taken this course during the last fifteen years have consistently indicated that a sound knowledge of basic developmental anatomy is required for an understanding of the theoretical material presented in lecture. We have also added experiments and laboratories involving living embryos to remind the student that embryos are more dynamic than stained tissues arranged on a microscope slide. Furthermore, these experiments hopefully illustrate classic questions of development and the approaches used to analyze them.

The laboratory provides numerous resources: microscopes and slides, living embryos, demonstrations, films, models, reference books, instructors, and this manual. We have purposely left to the student some of the responsibility of organizing and coordinating these resources and integrating the information they provide into a meaningful whole. This is the most important task in the course: it involves some hard work and, we hope, some thinking.

This is the second revision of a manual that began as a few separate "handouts" to supplement a published manual. It is still very much in the process of evolution. We would be grateful to the reader for criticisms and suggestions for its improvement.

We are grateful to a number of outstanding teaching assistants who dutifully read and reread these chapters. Special thanks are extended to Tuan Duong, Mark Reedy, Holly Johnson and Kirsten Hall for their thoughtful comments and editorial acumen.

P. B. Armstrong
C. A. Erickson
R. D. Grey
J. K. A. Metzler

I. USE OF THE NIKON COMPOUND MICROSCOPE: Köhler Illumination

To obtain the best resolution and contrast, careful attention must be paid to obtaining proper condition of illumination. The directions are specifically for a Nikon microscope, but they are applicable to most compound microscopes.

1. Focus the preparation in the microscope. Adjust the distance between the oculars using the lever beneath the eyepieces. If necessary adjust the focus of the eyepieces for your eyes. Focus the specimen in the right eyepiece, then adjust the focus of the left eyepiece until the specimen is in focus there. Adjust the amount of light with the knurled ring on the right side of the base of the microscope.

2. Close condenser diaphragm somewhat (using the lever mounted at the bottom of the condenser).

3. Close the field diaphragm by turning the knurled ring on the field diaphragm lens.

4. Rack the condenser up or down until the image of the field diaphragm is focused.

5. The light source in the Nikon microscope is pre-centered. If the spot of light from the light source is not in the center of the field, tell your teaching assistant so that he/she can adjust this.

6. Open the field diaphragm until its image just disappears from the view. If the field is not evenly illuminated, move the light bulb holder in or out slightly.

7. It may be necessary to repeat steps 1-4 each time magnification is changed.

8. Adjust contrast using the condenser diaphragm. This should be closed not more than 1/4 of its largest aperture. If the condenser diaphragm is closed too far, resolution is lost.

9. Avoid hitting the objectives with the microscope slide. Avoid touching any of the lenses. If any lens needs cleaning, use lens paper only. Dust specks in the field of view probably are on top of the field diaphragm lens.

II. DRAWINGS

During the course, you will be asked to make drawings of embryonic structures. A drawing is a convenient means of making notes of what you observe with your microscope. It is also an excellent tool for sharpening your observational skills. You are required to do the assigned drawings. Simple outline drawings are all that is required; stippling and shading are unnecessary. Label all the structures you can identify with the aid of this lab manual, other references provided in lab, and your TA. Labels should be as neat and as explicit as possible.

The purpose of the drawings is not to produce artistic creations, but rather to aid you in making accurate and careful observations. The drawings are to be handed in at the end of the laboratory period when the work is completed. Working on the drawings outside of lab, or touching them up, defeats their purpose.

The assigned drawings will be checked for accuracy by your TA and returned to you during the next laboratory session. Any errors noted should be corrected prior to turning in your notebook at the end of the quarter.

In addition to the assigned drawings, you will want to make quick sketches of your observations of living embryos, of organ systems as you study them, and as notes to help you remember what you have seen. As you look at your slides, learn to recognize tissue types. When you are trying to identify a structure think first, what type of tissue is it. Then note where it goes and to what it is connected as you follow the structure from section to section. Use landmarks to help find your way around the embryo. When a distinctive, easily recognized structure is associated with another less easily identified structure use the easy one as a landmark.

Introduction

The Vertebrate Body Plan

I. Introduction

This text focuses primarily on the development of the vertebrate body. To understand how the body of a complex organism such as yourself is constructed from a single-celled egg, it is necessary to understand something about the basic design of the finished product. Your task is therefore somewhat akin to that of an engineer who wants to know how a particular building was constructed. It is essential not only to become familiar with the individual components of the vertebrate body—organs, tissues, etc., but to understand also the architectural relationships of these parts in the integrated complex in which they exist and function.

Vertebrates comprise the subphylum **Vertebrata** belonging to the phylum **Chordata**, which also includes two invertebrate subphyla: **Urochordata** (which includes the tunicates, or sea squirts,) and **Cephalochordata** (which includes Amphioxus). As chordates, vertebrates possess the three structural features characteristic of the phylum: a single **dorsal hollow nerve tube**, a **notochord**, and an enlarged cranial end of the gut (the **pharynx**) with an associated **internal gill apparatus**.

Although these three features appear in all chordates at some stage during their embryonic development, some traits become so highly modified (or even lost entirely) as the embryo transforms into an adult that the basic similarities in the body plan are not always apparent in mature organisms. In some of the invertebrate chordates for example, the neural tube becomes vestigial after metamorphosis. And in most classes of vertebrates the notochord is no longer apparent in the adult stage, having been replaced by the vertebrae of the spinal column. Likewise, the developmental fate of the gill apparatus in the so-called "higher" vertebrates (reptiles, birds, mammals) is quite different from what occurs in "lower" aquatic forms (fish and amphibians). In the more advanced vertebrates the rudimentary gill apparatus that forms in the embryo is subsequently transformed into a variety of non-respiratory structures associated with the mouth and pharynx; in lower forms the gill rudiments develop into full-fledged gills used for respiration.

The embryos of different classes of vertebrates offer us a chance to examine the central themes on which the anatomy of all vertebrates is based. These similarities in design are most evident at the embryonic stages just after the body begins to take shape. The major elements of vertebrate architecture will be identified using the embryo of an amphibian. As you observe the organizational features of the embryo, pay special attention to the **relative** positions of the various body components. The liver, for example, lies posterior to the head and ventral to the gut. This spatial relationship holds true for all vertebrates, as do most of the other relationships you will observe in the amphibian embryo.

Several alternatives for study are presented in this chapter. The first part deals with major chordate and vertebrate features that can be identified in whole or intact embryos. Many of these features can be identified in living embryos and in embryos that have been commercially prepared as "whole mounts." The internal construction of amphibian embryos can be examined by studying commercially prepared slides of embryos that have been cut into thin sections and stained.

Embryos of the common leopard frog, *Rana pipiens*, are frequently used for commercially prepared whole mounts and cross sections. Two stages of development are commonly available, and are identified according to the length of the embryo: the 4 mm-stage and the 6 or 7 mm-stage. Although the descriptions in the text deal primarily with newly hatched (4-mm) *Rana pipiens* embryos, any of several other species (e.g., *Bombina orientalis*, *Xenopus laevis*, or *Ambystoma*) provides excellent laboratory material. When studying embryos of these other species, keep in mind that they may vary slightly from *Rana pipiens*; these variations must be taken into account when interpreting the descriptions provided in the text.

II. Architecture of Intact Embryos

A. Fixed Embryos

The amphibian embryo provides a superb example of the chordate body plan. At the early tailbud stage, many chordate and vertebrate features can be seen very simply by careful and direct observation of the embryo's external appearance. Use a large-mouth pipet to transfer one of the embryos to a small petri dish and examine the embryo with a dissecting microscope.

Good lighting or illumination is essential for the observations you are to make, so take time to adjust your light source properly. Adjust the illuminator so that light is cast obliquely across

the embryo, producing shadows that enhance contrast and permit you to see the contours of the body, as illustrated in Figure 1.1. Position the illuminator close to the specimen to obtain a maximum amount of light.

NOTE:

If the embryo has not yet hatched, it is necessary to remove the jelly layers. Using two pairs of watchmaker's forceps, grasp the jelly with one pair, and with the other, tear pieces of jelly from the embryo. Eventually, the embryo can be popped out of the fertilization envelope and the innermost layer of jelly.

Several vertebrate features are immediately apparent. The animal is obviously **bilaterally symmetrical**, and its anterior end is the site of specialized sense organs associated with the expanded neural tube. The expansion and specialization of the anterior portion of the neural tube into a brain and the development of specialized sense organs associated with the brain are typical vertebrate characteristics.

The sense organs associated with the brain, and the other major external features, can be seen by turning the animal on its side. The most obvious structures from this view are bulges representing the positions at which major organs are forming (see Figure 1.1). The developing **eye** is the most anterior of these major structures, and appears as a prominent bulge on the side of the head. In older embryos the eye may be lighter in color than other parts of the embryo. Note the location of the **pronephros** in Fig. 1.1, and identify the site of this structure on your own embryo. The pronephros is the first kidney or excretory organ to form in all vertebrate embryos. In lower vertebrates it functions throughout most of the larval period. Later the pronephros is superseded by a more posteriorly located kidney, the mesonephros.

The **pharynx**, or **gill region**, lies between the eye and the pronephros. In early embryos it appears simply as a large swelling, indicative of the lateral expansion of the gut tube at that site. In slightly older embryos, the bulge in the gill region may display linear gill clefts or depres-

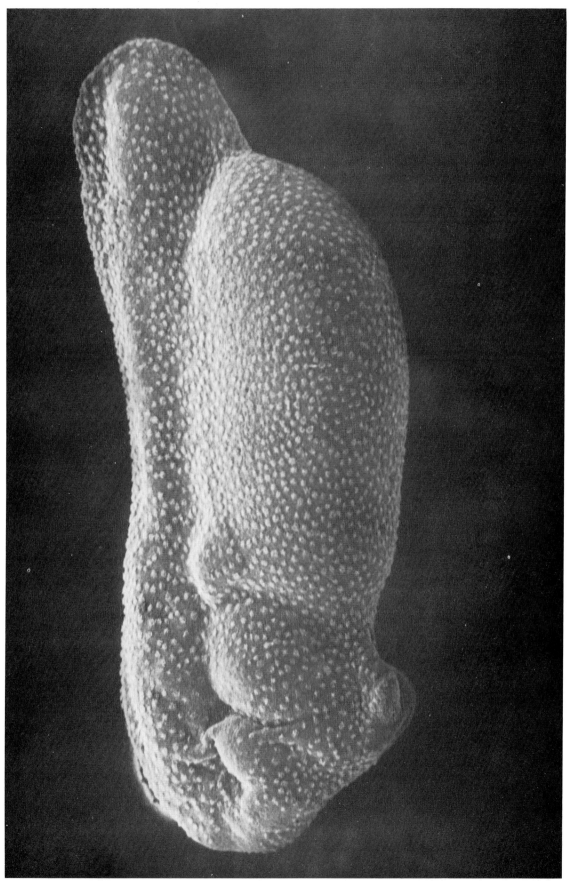

Figure 1.1. *Scanning electron micrograph of a* Rana pipiens *embryo.*

sions oriented at right angles to the body axis. By late tailbud stages, the gills appear as feathery projections from this region.

Finally, note that the future **nose** or olfactory organ is at the tip of the head, at about the same level as the gills. Just below the gills (in frog embryos) are the two **mucous glands** or suckers which the animal uses to attach to plants or other food material.

B. Whole Mounts and Mid-Sagittal Sections of *Rana pipiens* Embryos

Some of the major anatomical features of young amphibian embryos can be seen better in specimens that have been treated to permit observation of internal structures. In these specimens the lipids were extracted by soaking the embryos in an organic solvent such as toluene or xylene. The embryos were then stained and mounted intact on a microscope slide. The structures visible in these preparations are shown in Figures 1.2 and 1.4.

When examining whole mount preparations with a microscope, please note that **these preparations are thicker than the usual histological slides** and require special care in handling. The whole mounts can be best observed using a dissecting microscope with transmitted illumination.

NOTE:
If a compound microscope must be used, a 3.5X or 4X objective is best. Special caution is required to avoid crushing the coverslip when focusing or changing objectives.

1. EARLY TAILBUD (4-MM) STAGE

In the 4 mm-stage embryos (Figure 1.2) the most obvious structures are the following:
Eye. The wall of the optic vesicle forms a prominent circular profile in the head region.

Brain. The dorsal hollow neural tube, characteristic of chordates, is expanded in the anterior region, a feature that is most apparent in the area above the eye. Most internal structures are difficult to visualize in whole mounts, and it is therefore essential to examine these structures in

sagittal sections. Such sections are made by slicing the embryo in a plane parallel to the body axis. The contours of the brain and spinal cord are clearly seen in a mid-sagittal section through the 4-mm embryo (Figure 1.3). After studying this section, it is relatively easy to locate the outlines of the brain and spinal cord in the whole mount.

In all vertebrates, the brain has three principal subdivisions: the forebrain or **prosencephalon**, the midbrain or **mesencephalon** and the hindbrain or **rhombencephalon**. Identify these regions, first in the mid-sagittal section, then in the cleared whole mount. In the latter, notice the relatively thin roof of the rhombencephalon. This feature will be more obvious in older embryos.

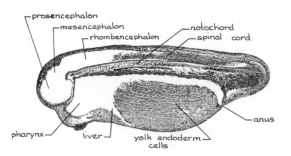

Figure 1.2. *Cleared whole mount of the newly hatched (4-mm body length)* Rana pipiens.

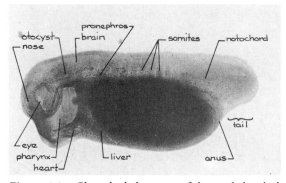

Figure 1.3. *Mid-sagittal section of a 4-mm embryo of* Rana pipiens. The three principal divisions of the brain (prosencephalon, mesencephalon, and rhombencephalon) are easily distinguished. This particular section grazed the lateral wall of the spinal cord, giving it a solid appearance in some regions. Note the organization of the gut tube with the broad, anterior pharynx, ventral liver diverticulum, and intestinal tube.

4

Pharynx. The pharynx is an anterior expansion of the gut which, in whole mounts, usually appears lighter in color than surrounding regions. The generous dimensions of the pharynx can be better appreciated by examining a sagittal section (Figure 1.3).

Liver. At this stage the future liver is a simple, finger-shaped hollow pouch that extends posteriorly and ventrally from the pharynx, easily identified in sagittal section (Figure 1.3). Because it is hollow, the liver, like the pharynx, appears lighter than the surrounding tissues in stained whole mounts.

Somites. The somites are blocks or segments of tissue, arranged in a row on each side of the spinal cord. Later in development somite cells give rise to the cartilage and musculature of the spinal column, and also contribute to the dermis, the deepest layer of the skin.

Anus. Bringing up the rear is the posterior opening of the gut, as identified in Figures 1.2 and 1.3.

2. YOUNG LARVAL (7-MM) STAGE (Figure 1.4)

By the 7-mm stage, the components identified previously have continued to develop, but can still be recognized from the earlier descriptions. Identify the eye, brain (which now has differentiated into distinct regions), somites, anus, and post-anal tail. Other vertebrate features, not previously described, are now apparent:

Nose (nasal pits). At this stage the nasal pits appear as two small, circular depressions or "pits" located anterior, and slightly ventral, to the eye. Cells comprising the pit later form the epithelial lining of the nose and generate the olfactory nerve, which connects the nose to the brain.

Otocysts (otic vesicles). The otocysts are spherical vesicles or sacs that lie alongside the brain in the region posterior to the eyes. They later become the inner ears.

Heart. The heart can be seen directly beneath the floor of the pharynx. Only the general outline of the heart can be discerned in whole mounts; its structure can be observed more clearly in cross sections.

Notochord. The notochord is a solid, flexible rod lying immediately beneath the spinal cord. It is the first, and in early stages, the only structural support of the embryo.

Kidney. In the larval amphibian, the functional kidney is the **pronephros** or "head" kidney. Although the pronephros may be difficult to observe in whole mounts, its contours can be observed immediately ventral to the third, fourth, and fifth somites (Figure 1.4). A duct (the **pronephric** or **Wolffian duct**) runs along the ventral border of the somites, connecting the pronephros to the widened posterior end of the gut tube or **cloaca** (Figure 1.4).

Somite. The paired blocks of tissue arranged in single rows along both sides of the spinal cord are the somites. These will later give rise to the cartilage of the vertebral column, as well as the muscles associated with it, and will contribute cells to the connective tissue layer of the skin (the dermis).

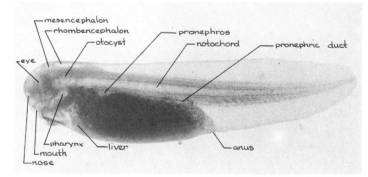

Figure 1.4. *Cleared whole mount of the 7-mm embryo of* Rana pipiens. In addition to its increased length, the embryo shows other developmental features not evident at earlier stages (e.g., Fig. 1.2). Gills have developed as feathery lateral processes protruding from each side of the pharynx. The special sense organs (nose, eyes, ears) are more prominent, as is the principal excretory organ, the pronephros.

5

Endoderm. This is the extensive mass of tissue composed of large cells lying ventral to the somites, and forms the primordium for the gut and the digestive glands. These cells contain large quantities of yolk, and in life are yellow colored.

Intermediate mesoderm. The narrow band or strip of tissue lying between the somites and the endoderm is the intermediate mesoderm. At its anterior end, it is thickened into the primordium for the larval kidney or pronephros. The pronephros appears as an egg-shaped bulge lying beneath somites 3-5.

Brain. The anterior expansion of the neural tube becomes the brain. The bulges from either side of the brain are the developing optic cups, which are the primordia for the retina of the eye and the optic nerve.

Pharyngeal region. The expanded region of the gut lying between the brain and the pronephros will form the pharynx and associated structures. The lateral walls of the pharynx are composed of **pharyngeal pouches**, lateral outpocketings of the gut in this region, and **visceral arches**, which are the columns of tissue lying between the pouches (see Figure 1.5).

C. Free-Hand Sectioning of Intact Fixed Embryos

Embryos that have been fixed in formaldehyde or, better, in glutaraldehyde can be cut cleanly using a fragment of a razor blade. Place an embryo in a small petri dish or similar container containing several milliliters of tap water and cut it with the blade. Move the blade through the embryo with a single slicing motion; do not "saw" back and forth or the tissue will crumble.

Make one cut transverse to the main body axis in the gill region and a second cut through the trunk region. The central block of the embryo can then be stood on end in the dish and many features of its internal anatomy can be observed. Particularly evident is the dorsal **neural tube**, flanked by **somites**. The **notochord** lies just beneath the neural tube. Try to dissect out a piece of the notochord. How far does it extend

anteriorly? The gut tube lies beneath the notochord. Notice how much wider the gut is in the pharyngeal region.

D. Representative Cross Sections of a 6- to 7-mm Frog Embryo

Now that the general organization of the embryo has been studied a further appreciation of its anatomy can be gained by examining prepared slides that show features not visible by gross inspection. On the slides are very thin sections or slices that were cut in a plane approximately perpendicular to the long axis of the embryo's body. Since the embryo is usually curved slightly, the plane of sectioning is not perfectly perpendicular, and the sections on the slide are often asymmetrical as a result. It is possible to section a complete embryo from head to tail. Starting with the head region, all the sections are laid on a slide from left to right, just like the printed words on this page. When all sections from a single animal or specimen are mounted in this way, the preparation is called a set of **serial sections**. For the study of the frog embryo, only sections through certain selected body regions have been mounted on the slide; these preparations are called **representative cross sections**.

1. PRELIMINARY SURVEY

Begin with the first section on the left end of the top row of the slide using the lowest power of the microscope. Examine the first section, which is from the anterior region of the embryo, and then quickly scan the other sections on the slide, reading from left to right. Remember that the image in the compound microscope is reversed: this reversal makes structures on the right of the embryo appear on the left, and also accounts for the apparent reversal in direction of movement as you shift from section to section. The sections have been stained so that the tissue components are visible in the light microscope. The tissues are usually stained with two dyes, a nuclear stain such as hematoxylin (blue), and cytoplasmic stain such as eosin (reddish orange). There may be cracks or broken areas of some sections. Because they contain large amounts of yolk, embryonic amphibian tissues often

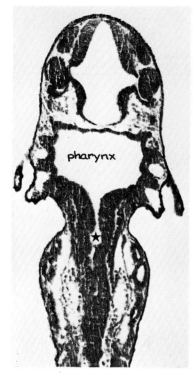

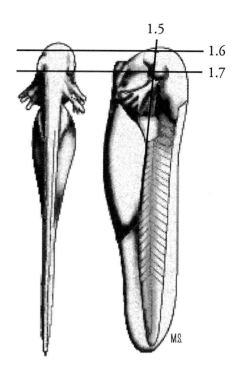

Figure 1.5. *Light micrograph of a frontal section through the pharyngeal region of a 7 mm-stage (young larval) frog embryo.* The endoderm lining the lateral walls of the pharynx bears a series of outpockets called pharyngeal pouches. These pouches are characteristic of all chordates. The structure over the pharynx is the diencephalon, with optic cups at each side. Posterior to the pharynx, the gut tube narrows. The liver diverticulum is indicated by a star.

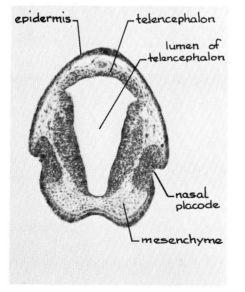

Figure 1.6 *Nasal placode level.*

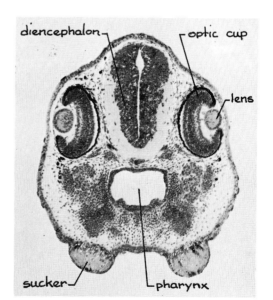

Figure 1.7 *Eye level.*

7

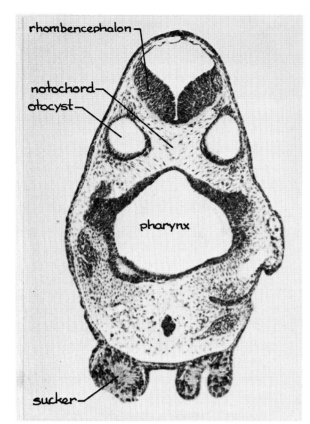

Figure 1.8. *Ear Level.*

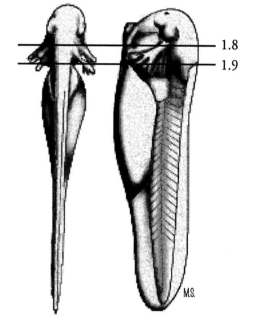

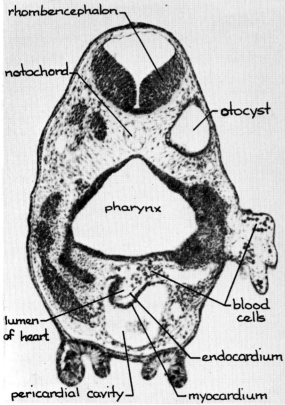

Figure 1.9 *Heart level.*

8

become brittle and crack as they are prepared to be mounted on slides.

2. SIX MAJOR REGIONS OF THE EMBRYO: HOW TO PROCEED

Examine in detail the six levels or regions described below, proceeding as follows. Figures 1.6 through 1.11 are photomicrographs of each level or region of the embryo to be studied. Begin by locating the section on your own slide that best corresponds to the photomicrograph of the region being described. Then, make a simple line drawing of that section in your notebook. Identify and label all major structures (**boldface** in the text) described for each region. Keep in mind that the sections you are studying are only representative. Hence, all sections on your particular slide are not likely to match exactly all of the descriptions that follow. If your slide lacks one or more of the regions described, you should either borrow a slide from a neighbor or use the slides on demonstration.

Finally, as you study each level, refer to the whole-mounted diagram handout (or Figure 1.5), and draw a line through the embryo indicating the plane of the sections on your slide.

Nasal placode region (Figure 1.6). The compact layer of darkly stained tissue around the perimeter of the section is the **epidermis**, which covers the body, and is only one or two cell layers thick. Under high magnification the brownish pigment granules that give the embryo its color can usually be seen in these cells. The **nasal placodes** appear as thickenings in the epidermis on the lateral surface of the head. Your section may show one or both of these thickened regions invaginating to form the sensory epithelial tissue that will line the nose. The prosencephalon (forebrain) by now is divided into two regions, the **telencephalon** (the anterior-most segment, associated with the nasal placodes) and the **diencephalon** (more posterior and associated with the optic cups). The space in the middle of the section is the lumen or cavity of the **telencephalon**. In the living embryo, this cavity is filled with fluid. The appearance of the brain (particularly the shape of the lumen) will vary, depending upon the exact region from which your particular section was cut. Usually, the wall of the brain in this region is rather thick on the sides, and thin at the narrow dorsal and ventral regions. The tissue of the brain walls is epithelial at this stage of development (**neural ectoderm**). This tissue will later differentiate into the neural and connective tissues of the brain. Notice the loosely arranged cells lying between the walls of the brain and the skin. These are **mesenchyme** cells and are involved in the formation of supporting and connecting tissues of the head (e.g., bone, muscle).

The eye level (Figure 1.7). This section passes through the "C"-shaped **optic cups**, which lie just below the epidermis on each side of the section. The **lens** of the eye appears as an oval structure, filling the gap in the "C". In vertebrates, the eyes originate as bulges that push laterally from the side-walls of the forebrain. The lenses form from ectoderm immediately overlying the developing optic cups. Between the two optic cups is the **diencephalon**, the region of the forebrain from which the optic cups have originated. Each optic cup is joined to the brain by an optic stalk, which may not appear in your section. The cavity just below the diencephalon is the **pharynx**, the anterior region of the gut tube.

The ear level (Figure 1.8). The brain is the most prominent structure in the upper part of this section. This region of the brain is called the **rhombencephalon**. The rhombencephalon can be easily recognized because of the extremely thin roof that covers the Y-shaped brain cavity. The two hollow **otocysts** (**otic vesicles**) lie just below the rhombencephalon. The otocysts first appeared at an earlier stage of development as thickenings or placodes in the epidermis, similar in appearance to the nasal placodes. The vesicles formed when the flat placodes invaginated and became first cup-shaped and then spherical as the rims of the cups joined and fused. Later, the otocyst will give rise to the inner ear of the frog. The **notochord**, which lies just below the rhombencephalon, appears as a circular profile consisting of large, vacuolated cells surrounded by a thin, darkly stained sheath of non-cellular material. In the living animal, the vacuoles in these cells are filled with fluid, making the noto-

9

chord a turgid, but flexible, supporting structure. The **suckers** can usually be seen on the ventral surface of the body in these sections.

The heart level (Figure 1.9).

In this section, the rhombencephalon and notochord appear much the same as at the preceding level. Lying below the notochord, the thick-walled structure with a lumen is the gut. The **heart** lies just below the gut in the **pericardial cavity**. The appearance of the heart varies according to the level from which the section was taken. Usually, two layers of the heart wall can be seen: the thin inner lining is the **endocardium**, the thicker, outer wall is the **myocardium**, or muscular layer of the heart.

What are the structures projecting from the body that can often be seen in sections from the ear or heart levels?

Pronephros region (Figure 1.10). Identify the neural tube and the notochord. Flanking the notochord are the somites, seen earlier in the whole-mount preparations or the living specimens. They appear as dense concentrations of mesenchymal cells. The **gut** appears as a thick-walled structure comprised of darkly stained cells, lying below the notochord. On some slides, the lumen of the gut appears as a long slit that projects ventrally. Any outpocketing of the lumen is called a **diverticulum**; this particular diverticulum develops into the **liver** at a later stage.

The **pronephros**, which you saw in the living embryos, or whole mounts, can be identified as the clusters of ring-like structures lying on either side of the dorsal surface of the gut. These circular structures are actually **pronephric tubules** cut in cross section. The mass of cells surrounding the gut and filling the ventral part of the section is the yolk-filled endoderm seen earlier in living animals and whole mounts.

Sections through the mid-gut (Figure 1.11). The lumen of the neural tube is considerably narrower in this section; this feature indicates that the section passes though the **spinal cord**. Identify the notochord and somites. The pronephric ducts lie just below the ventro-lateral regions of the somites. Just beneath the notochord, your section may show a small darkly stained structure, the subnotochordal rod, the origin and function of which is unknown. The circular space beneath this rod is the **dorsal aorta**, the major vessel that transports blood to the body regions posterior to the heart. You may be able to identify blood cells lying in the aorta. Ventral and lateral to the somites, identify the **pronephric duct**. This transports urine produced by the pronephros to the cloaca.

The ventral half of the body in this section consists of the large yolk-filled endoderm cells. You may be able to find the lumen of the gut in this mass of cells; it appears as a cavity lying ventral to the dorsal aorta.

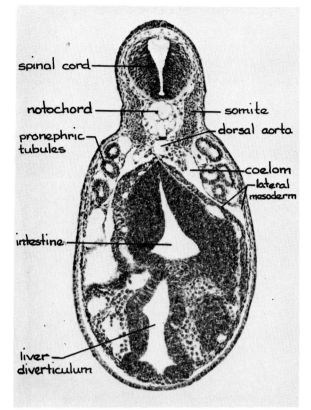

Figure 1.10 *Pronephros region.*

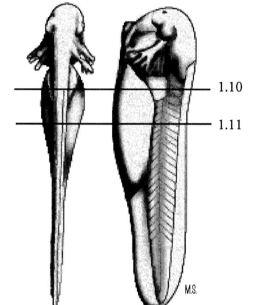

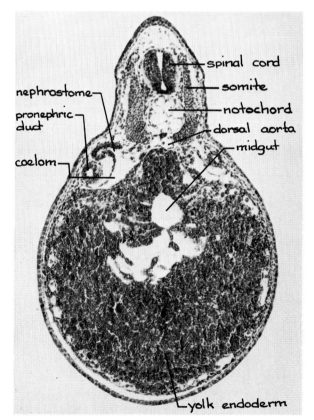

Figure 1.11 *Midgut region.*

zygote with an endogenous energy reserve in the form of yolk. The yolk serves to "feed" the developing embryo until it reaches a stage at which it can utilize food from external sources.

The principal cellular contribution of the male gamete, the **spermatozoan** or **sperm**, is the nucleus, which carries the haploid genome derived from the father. Few cytoplasmic components are added to the zygote by the sperm, with the exception of the mitochondria and centrioles, which in most species enter along with the nucleus.

As knowledge about the process has increased, the term "fertilization" has been expanded to comprise a family of events, all of which center on the union of gametes. The general order of these events is outlined below; variations in the exact sequence are seen in some species:

1) **Activation of the sperm**, which occurs in most animals in response to oviducal secretions or to jelly coats surrounding the ova. In mammals, the activation of sperm is called "capacitation," and occurs in the reproductive tract of the female.

2) **The acrosome reaction**, which may occur as the sperm makes its way through the egg integuments or may be induced by the egg investments after the sperm is bound to them. The acrosome reaction is a necessary prerequisite to successful sperm-egg fusion in those organisms in which it occurs. These include echinoderms, coelenterates, some fishes, amphibians, mollusks and mammals.

3) **Penetration of the sperm** through the external coats or envelopes that surround the egg.

4) **Fusion of the sperm with the egg**, i.e., fusion of the plasma membranes of

Fertilization

I. Introduction

Fertilization is the point in the life cycle of a sexually reproducing species that marks the beginning of a new individual. Fusion with the sperm releases the egg from its arrested state of development and metabolism, and sets in motion the train of events that culminate in the formation of a sexually mature individual resembling the parents who produced the gametes. It is also at this point in the life cycle that the diploid condition is restored.

The contributions of the male and female gamete to the newly formed zygote are very different. The female gamete, or **ovum**, provides not only a haploid set of chromosomes, but also the cytoplasm required for formation of all cells up to the larval stage. The egg's cytoplasm was, of course, formed in the ovary, under the direction of the maternal genome, and contains the enzymes and metabolic components required to drive the early events of development. The cytoplasm of the egg is also known to contain genetic information, in the form of stored messenger RNA, that directs the synthesis of some of the proteins required for early development. In addition, the ovum also provides the newly formed the two gametes.

5) Establishment in the egg of blocks against entry of additional sperm, i.e., **the blocks to polyspermy.**

6) **Completion of meiosis.**

7. **Entry of the sperm into the egg** cytoplasm and the subsequent union of the haploid nuclei (pronuclei) of the two gametes to form the diploid nucleus of the zygote.

8. **Metabolic activation** of the egg, and the initiation of development. The known order of metabolic events in sea urchin embryos is shown in Figure 2.1.

Sea urchins have been long-standing benefactors to those biologists interested in fertilization. Their gametes, which are produced in great abundance, can be maintained and observed in sea water, the natural medium in which fertiliza-

Table 2.1: Biochemical Events Common to Sea Urchin and Mammalian Fertilization.

1. The sperm membranes contain species-specific proteins involved in "recognition" of the egg surface before the acrosome reaction

2. Occurrence of an acrosome reaction

3. Calcium is necessary for the induction of the acrosome reaction

4. Cyclic nucleotide metabolism is altered in sperm during the acrosome reaction

5. A protease is exposed as a result of the acrosome reaction

6. Sperm binding is to receptors on the extracellular egg coat: the vitelline layer in sea urchins and the zona pellucida in mammals

7. Attachment of sperm to eggs after the acrosome reaction exhibits species-specificity

8. "Sperm receptor" glycoproteins can be isolated from both the zona pellucida and the vitelline layer

9. A calcium transient in the egg triggers the cortical granule reaction after sperm-egg fusion has occurred

10. A trypsin-like protease released from cortical granules destroys the sperm-binding capacity of the sea urchin vitelline layer and the mammalian zona pellucida

11. A peroxidase released from cortical granules hardens the vitelline layer and the zona pellucida by catalyzing the formation of di- and tri-tyrosine cross-links between the component proteins

12. Fusion of sperm and egg results in activation of the new synthetic machinery of the egg and in cell division

13. Sea urchin and mammalian sperm have cross-reacting surface antigens

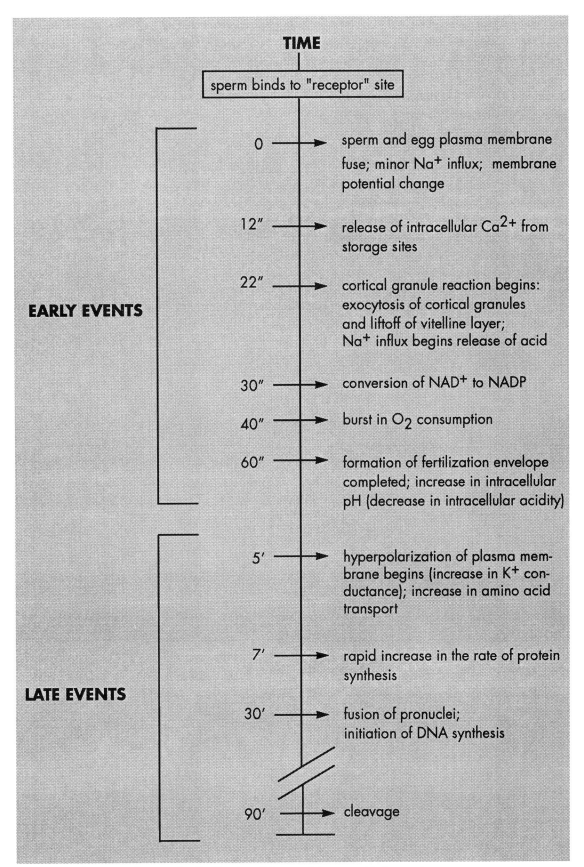

TIME

sperm binds to "receptor" site

EARLY EVENTS

0 → sperm and egg plasma membrane fuse; minor Na^+ influx; membrane potential change

12″ → release of intracellular Ca^{2+} from storage sites

22″ → cortical granule reaction begins: exocytosis of cortical granules and liftoff of vitelline layer; Na^+ influx begins release of acid

30″ → conversion of NAD^+ to NADP

40″ → burst in O_2 consumption

60″ → formation of fertilization envelope completed; increase in intracellular pH (decrease in intracellular acidity)

LATE EVENTS

5′ → hyperpolarization of plasma membrane begins (increase in K^+ conductance); increase in amino acid transport

7′ → rapid increase in the rate of protein synthesis

30′ → fusion of pronuclei; initiation of DNA synthesis

90′ → cleavage

Figure 2.1. *Sequence of events after fertilization of the egg of the sea urchin* Strongylocentrotus pupuratus.

tion and development occur. Also, the eggs of sea urchins tend to be among nature's largest, making direct microscopic observation of gamete interaction a relatively simple matter. In addition, sea urchin and mammalian fertilization have many biochemical features in common, making sea urchins a useful model system for the study of the molecular mechanisms of fertilization (Table 2.1).

Eggs of sea urchins differ from those of vertebrates in that the elevation of the vitelline envelope is a highly exaggerated affair. Unlike the situation in many vertebrates in which the unfertilized egg is surrounded by a distinct, capsule-like vitelline envelope, the unfertilized sea urchin egg has only a thin (100-200Å) layer of glycoprotein deposited on the surface of its plasma membrane. Presumably homologous to the thicker vitelline envelope of other egg types, this material is sometimes called simply the vitelline layer. Starting at the point of sperm entry, the vitelline layer, which is tightly apposed to the plasma membrane, lifts off from the egg's surface like a rapidly inflating balloon and forms the fertilization envelope. As the vitelline layer lifts off, sperm bound to it detach. This is owing to removal of sperm "receptors" from the vitelline layer-fertilization envelope by a protease released from the egg shortly after sperm-egg fusion.

One of the aims of this exercise is to provide you with the opportunity to observe living gametes. In dealing with the phenomenon of fertilization, it is important for you to see the gametes, to observe their interactions, and to enjoy one of nature's most dramatic and fascinating performances. The other aim of the chapter is for you to design experiments that test some of the hypotheses that have been formulated to explain some of the mechanisms of gamete interactions.

II. Observation of Gametes and Their Interaction

Simple observation is a critical part of any science, but it is a component with which many beginning investigators tend to become impatient and to neglect. In guiding

your observations of gametes, the manual will point out various features that are worthy of your attention. You should not, however, limit your observations to these guidelines. One danger in providing guidelines is that you tend not to see anything else and thereby fail to discover something for yourself that causes you to wonder. There is no known recipe for success in the process of discovery by observation, but two essential ingredients are curiosity and patience, applied in generous quantities. New discoveries are usually made when a curious-minded observer sees and wonders about something that previous observers have overlooked.

One time-tested method to enhance effective observation is to make simple drawings or sketches of what is observed, and we encourage you to do this in your notebook. Some of the important features to keep in mind and to note in your records are the following:
1) Size relationships. For example, how big is the sperm *relative* to the egg?
2) Spatial relationships. *Where* is one component or object in relation to the others in the system?
3) Behavior.
4) Sequence of events. In what *order* do various interactions occur?

A. Sperm and Egg Preparations
1. OBSERVATION OF SPERM

READ THIS BEFORE PROCEEDING!!!!!

1) It is extremely important that you keep sperm and eggs separate until you are ready to fertilize. The tiniest drop of sperm can ruin a beaker full of eggs by fertilizing them at the wrong time. The sperm and eggs should be kept on opposite ends of the bench (your lab instructor will show you where they are). **DO NOT** use the same pipette, beaker, or test tube for both sperm and eggs.

2) Use disposable plasticware whenever possible. Even clean glassware may be contaminated with substances toxic to sperm and eggs.

The laboratory instructor will prepare a dilute sperm suspension by adding 1-2 drops of "dry" sperm to 10 ml sea water. ("Dry sperm" refers to undiluted semen, i.e., the sperm as it comes out of the sea urchin. Dry sperm is about 80% immotile, anoxic sperm, and 20% seminal fluid. Upon dilution, the sperm become motile, and capable of fertilizing eggs.) The clumped, dry sperm are suspended by pipetting them a few times with a disposable pipette. Put one drop of the sperm suspension on a slide, add a coverslip, and observe the preparation under high power (400X). If you have trouble seeing the sperm,

NOTE:

DO NOT USE THE HIGHER MAGNIFICATION OBJECTIVES UNLESS YOUR PREPARATION HAS A COVERSLIP.

move the condenser down and close the iris diaphragm. Note the swimming motion of the sperm. Some may have undergone the acrosome reaction on contact with the glass and will be firmly bound to the glass by the tip of their head region. This will be particularly true if you are using *Lytechinus pictus* sperm.

2. OBSERVATION OF EGGS
Since the eggs will be crushed by a coverslip that is placed directly on the slide, use "depression" slides if available. If not, it is necessary to support the coverslip on solid "feet" of some sort. A simple way to accomplish this is to use tiny bits of modeling clay scraped from a ball of clay. Bits of clay under each corner will support the coverslip above the slide, preventing crushing of the eggs. (The clay "feet" should be quite small, e.g., about 1/4 mm in diameter, since the sea urchin egg is only about 80 μm or 0.08 mm in diameter.) If clay is not available, the coverslip can be supported with fragments of a broken coverslip, as follows: Break up a coverslip and place a few fragments of glass on a clean slide. Add a drop or two of eggs in sea water and cover

with a clean coverslip. The glass fragments act as spacers, which protect the eggs from being squashed. In some cases, visibility is improved if contrast is increased by underfocusing the condenser a little below the point where the stage diaphragm is in focus.

The jelly coat surrounding the egg is difficult to observe under normal conditions because its refractive index is nearly identical to that of sea water. You should be able to get an idea of the size of the jelly coat, however, because it prevents the eggs from coming in direct contact with each other by forming a spherical cushion around each egg. When the egg is freshly ovulated the jelly coat is smaller; as the egg remains in suspension for a period of time (or is washed several times in fresh sea water, as your eggs have been) the jelly coat hydrates and expands. If dejellied eggs are available, put a drop on a slide and observe. Since the jelly coats have been removed, you should be able to see eggs that are in very close contact with each other. Is it possible to fertilize dejellied eggs? Why or why not? Test the hypothesis.

You will occasionally note jelly layers that have come off the eggs. Watch how sperm agglutinate or "swarm" when they come in contact with the jelly. Jelly-induced sperm agglutination will spontaneously reverse itself and the sperm will swim away after a few minutes.

Take a clean slide and on it place two drops of egg jelly solution. Now add a drop of sperm, and observe their behavior. The sperm should form clumps ranging in size from about 200 μm to several millimeters in diameter. The size of the clumps will depend on how concentrated the jelly solution is. Little is known about this response of sperm. It appears to be independent of the acrosome reaction, a necessary step for successful fertilization. In some species, such as *Lytechinus pictus*, the swarming response does not occur at all, suggesting it is not an absolute prerequisite for sea urchin fertilization.

B. Fertilization
Take a drop of eggs and add a drop of the dilute sperm suspension. Remember not to touch any-

thing that has sperm on it to the stock egg suspension. Observe the binding of sperm to the egg surface. Several thousand sperm may bind to a single egg. However, usually only one sperm fuses with the egg. See if you can see the fertilizing sperm. It may be distinguished from the other sperm because it stops moving sometime around the time of sperm-egg fusion. The vitelline layer will begin elevating at the site of gamete membrane fusion. Use your watch to time how many seconds elapse from sperm addition to the first visual elevation. How much time is required for complete elevation? Repeat the observation five times. Record your observations in your notebook and determine the average time for the beginning of the liftoff and the completion of the liftoff.

After the elevation of the fertilization envelope has been completed, add a fresh drop of sperm suspension. Do sperm bind to the elevated envelope?

Within two or three minutes you should be able to see the hyaline layer as a thin, clear layer bound to the plasma membrane of the egg. To see it you may need to close the iris diaphragm and move the condenser down. During cleavage the hyaline layer remains bound to the surface of the outermost blastomeres. You will observe the hyaline layer in more detail in your study of early embryonic stages (Chapter 3).

What other features of gamete interaction were you able to observe? Make note of your observations in your laboratory notebook.

III. Experimental Analysis of Fertilization

A. Preparation and Planning

After you have made some observations on the behavior of sperm and on the responses of the eggs, and have gotten the knack of handling the material, you should then turn to an experimental analysis of some of the events you have observed.

The procedure for the experimental part of the laboratory will vary from the usual format in which the experiments are outlined for you. Instead, we have listed several problems that can serve as a starting point for your experiments. We also list and provide supplies and suggested conditions or procedures. But in each case, you and your lab partner are to design and execute the experiments.

You should work in pairs to conserve the materials. Design your experiments together— exchange ideas. Write down the step-wise procedure you intend to follow. Be sure to state clearly the question you are asking. A suggested organized method is as follows:
1) What question is the experiment designed to answer? The best experiments are those that can give clear-cut answers to sharply focused questions.
2) How are you going to observe or measure the result(s) of the experiment? What sorts of information will you need to record?
3) How many replicates of the experiment are required to give you reliable data?
4) What are the alternative results you might expect to obtain?
5) What are the necessary controls?
6) Remember to keep an accurate and detailed record of your experiments. The aim is not to produce a neat report of what you intended to do, but an accurate chronicle of what you actually did do.
7) After you have designed your experiments, you may want to check over your planned procedures with your laboratory instructor before proceeding. After completing your experiments, summarize them by stating the hypothesis to be tested by the experiment, outline the rationale and the design of the experiment, summarize the data collected, and formulate the conclusions. PLEASE NOTE: Because of the time limitations, each team will probably have time for only one or two experiments. Be sure to leave yourself time to record your data and, if possible, to observe the results of other teams who performed different experiments. Analysis of the data can be performed at a later time.

B. Problems for Experimental Analysis

NOTES ON PROCEDURE:

Traces of formaldehyde are instant death to gametes and embryos. Once formaldehyde contacts glassware, it cannot be removed. Therefore, use only designated glassware and disposable plasticware for formaldehyde experiments.

Treatment of gametes with various reagents should be done in small 5 ml test tubes. The results can be evaluated by taking samples from these reaction tubes and viewing them with a microscope; evaporation and heat from the microscope lamp will usually inhibit normal activity (e.g., sperm motility; fertilization).

Experiment 1. Qualitative assessment of sperm binding.

A time-course study of sperm-egg binding can be developed by fixing eggs in formaldehyde at varying times after fertilization. Fixation stops the fertilization process very quickly and preserves the egg with bound sperm.

    a. Materials: One member of the team should act as "timer" using the sweep-second hand of a watch. The other partner should act as "fertilizer." You will be provided with the following supplies:

        1) Living eggs and sperm.

        2) Four disposable 15ml test tubes for collecting samples of eggs at various times after insemination. Label the tubes in advance with the times (in seconds) after insemination at which samples are to be collected. Suggested times: 10, 25, 35, and 60 seconds.

        3) Formaldehyde-sea water fixative. Add 1 ml of fixative to each tube prior to the start of the experiment. (This fixative is made by mixing 15 ml concentrated formaldehyde with 85 ml sea water).

    b. Suggested procedure: At the command from the "timer," the "fertilizer" should add sperm to a suspension of eggs. Withdraw a pipetteful of eggs from this suspension at each of the designated times after addition of sperm. Add each pipetteful of eggs to the appropriate test tube (to which fixative was previously added).

> Be careful not to contaminate the pipette with formaldehyde during this step since formaldehyde reintroduced into the stock suspension of living eggs will kill them.

    c. Observations: Allow the eggs to settle to the bottom of the formaldehyde-containing tubes. Then remove as many eggs as you can in a small volume of fluid and place this concentrated suspension on a microscope slide. Add a coverslip provided with "feet" or spacers if you are not using a depression slide (See Section II, A, 2, above). Using drawings, record the abundance and position of bound sperm, and the progress of elevation of the fertilization envelope for each sample. What (if any) is the relationship between elevation of the fertilization envelope and the presence of bound sperm?

Experiment 2. Effect of trypsin inhibitor on the egg's response to fertilization.

a. Materials available:
1) Soybean trypsin inhibitor (SBTI).
2) Living eggs and sperm.
3) Formaldehyde-sea water fixative.
4) 5 ml test tubes for collecting fixed samples of fertilized eggs. Prior to beginning the experiment, add 1 ml of fixative to each tube. Label the tubes according to the times after insemination when samples are to be collected. Suggested times: 15, 30, 60, and 90 seconds (be sure to include a 90-second sample).

b. Suggested procedures:
1) Place about 1 ml of egg suspension in a disposable tube, allow the eggs to settle, and carefully remove most of the sea water above them. Replace with the SBTI-containing sea water (0.8 mg SBTI/ml sea water). Treat with SBTI for 30 minutes (on ice) prior to addition of sperm.
2) Fertilize eggs with sperm in a volume of sea water equal to the volume of SBTI-containing sea water in which the eggs are suspended (final concentration of SBTI = 0.40 mg/ml).
3) Fix the eggs at the designated times by adding an aliquot of the eggs to each of the tubes containing fixative.
4) Examine the eggs. How has treatment with SBTI altered the binding of sperm to the eggs? What effects do trypsin inhibitors have on the elevation of the fertilization envelope?

c. Think about controls for your experiment. What effects do you need to be aware of? Discuss appropriate controls with your lab instructor.

Experiment 3. Involvement of extracellular Ca^{2+} in fertilization

a. Materials available:
1) Living eggs and sperm
2) Calcium-free sea water (CaFSW)

b. Suggested procedures:
1) Wash a small amount of eggs with 4 ml CaFSW two times by centrifuging gently (200-300 rpm, 1 minute) using a tabletop centrifuge.
2) Resuspend the eggs in about 2 ml CaFSW.
3) Add a tiny amount of "dry" sperm and break up by gently pipetting into the egg suspension.
4) Look at the contents of the test tube and record your observations. Do the sperm "swarm" in the egg jelly coats? Do they bind to the egg vitelline layer? What do the elevated fertilization envelopes look like? Record your observations and draw what you see in your lab notebook.
5) Can you think of any other simple experiments to test the role of Ca^{2+} in fertilization? Try them.

Experiment 4. Species specificity of fertilization

a. Materials available:
  1) Living eggs and sperm of *S. purpuratus*.
  2) Living eggs and sperm of *L pictus*.
b. Suggested procedures:
  1) Obtain a small quantity of eggs and dry sperm of both species in test tubes; keep on ice.
  2) Examine each type of egg first alone, so they can be recognized; make sure that they are both dilute enough to see fewer than 50 eggs in a microscope field (about 2 drops of concentrated eggs/4 ml SW).
  3) Prepare a depression slide containing both kinds of eggs (about 3 drops of each dilute suspension). Check to make sure that you can see roughly equal numbers of each type of egg in a microscope field. Adjust the concentration if necessary.
  4) Prepare a small test tube containing the same concentration of eggs, but of larger volume (note the volume, in drops).
  5) Prepare a dilute sperm suspension of one of the types of sperm: 1 drop dry sperm/5 ml. Activate sperm by vigorous pipetting.
  6) Add about 2 drops of sperm for every 6 drops of eggs to the test tube and put it back on ice. In the depression slide, add 2 drops of the same sperm suspension and watch to determine the time of the first liftoff (and egg type) and the time that maximum liftoff occurs. Do NOT keep strong light on slide, or the eggs will die.
  7) After maximum liftoff, examine a sample from the test tube under the microscope. Examine several fields at random (select the fields without looking through the microscope), and record the number of eggs of each species that have liftoff. Count 3 to 5 fields of 20-50 eggs (so that you have counted at least 100 in different areas of the slide) and average the results.
  8) Repeat the experiment using a higher concentration of sperm (5 drops/5 ml) and record the results.
  9) Repeat the same experiment with the other species of sperm.

Experiment 5. Can extracellular Ca^{2+} activate eggs?

a. Materials available:
1) Living gametes
2) A23187 ($\sim$ 20 µM)
3) Calcium-free sea water (CaFSW)

b. Suggested procedures:
1) Pipette a small amount of eggs into each of two 5 ml conical test tubes. Label the tubes "normal SW" and "CaFSW". Add enough sea water to fill each to the 3 ml mark.

2) Wash the eggs in the tube labeled "CaFSW" twice with calcium-free sea water. Resuspend in 3 ml CaFSW.

3) To each test tube add one drop of A23187 and pipette gently to mix.

4) Is there a difference in the extent of activation of each set of eggs? From your observations, what can you say about the role of calcium on fertilization and activation? What do you think is the source of Ca^{2+} during normal activation and in this experiment? What evidence supports your ideas?

5) Can you think of any other experiment that will test the role of Ca^{2+} on the activation of development?

Early Development of the Embryo: Zygote to Gastrula and Beyond

I. Introduction

The morphogenetic events that follow fertilization are largely concerned with the re-packaging and rearrangement of the cytoplasm that is contained in the spherical zygote. During cleavage the cytoplasm of the large single-celled zygote is partitioned into many smaller undifferentiated cells called blastomeres. By the blastula stage, the blastomeres are positioned around a fluid-filled cavity and, unlike the solid, spherical configuration of the zygote, are amenable to the rearrangements and translocations that are to follow.

The first major rearrangement of cells occurs at the gastrula stage, and establishes in the embryo the "tube-within-a tube" configuration around which the mature body is fashioned. The gastrulation process also establishes the germ layers. These will form the primordia for all organs and tissues of the adult body.

The aim of this chapter is to examine the major themes and important events of cleavage and gastrulation. Observation of young sea urchin embryos allows us a glimpse of these processes in their simplest, least modified form. In the amphibians, a few modifications of these processes can be discerned. In addition to the opportunity to witness the intrinsic beauty of cleavage and gastrulation, observation of these embryos provides a perspective that will prove valuable for our later consideration of comparable events in birds and mammals.

The principal modifier of cleavage patterns and gastrulation movements is the bulky yolk stored in the egg. Yolk is a nutrient supply and energy reserve, literally the raw material and fuel required to construct the embryo. The degree to which yolk influences early morphogenesis is proportional to the amount of this reserve material in the egg.

The amount of yolk stored in eggs varies according to the adaptive strategy of each species or group. Different strategies are observed in the animal groups considered in this chapter. In the first, the sea urchin, the female spawns millions of eggs that are small and contain little yolk (**alecithal**, or **microlecithal**, eggs). Early development in these animals proceeds swiftly to the pluteus larva, a free-swimming, feeding stage. The amphibians we shall examine produce smaller numbers of eggs, but these eggs are about three

23

orders of magnitude larger (by volume) than those of sea urchins, and contain a significantly larger amount of yolk (mesolecithal eggs). While their early developmental period is, in general, more leisurely than that of sea urchins, they, too, develop into a larval form, the tadpole, that forages food for a time before continuing development to the adult form.

The amniote egg (birds) is large, laden with yolk (macrolecithal eggs) and the embryo develops directly into a miniature adult form.

II. Sea Urchin Embryos

A. Five Key Stages of Early Development

Living embryos of available stages will be provided by your lab instructor. (If living embryos are unavailable, the descriptions in this chapter may be used in conjunction with commercially prepared slides of either sea urchin or starfish embryos which have been preserved and stained. Development of starfish embryos generally follows the descriptions provided for sea urchins, but major differences will be noted where they occur). Examine one stage at a time. Use a pipet or dropper to remove a small sample from the embryo cultures and place a drop or two of the suspension on a clean microscope slide. **To avoid crushing the embryos, use depression slides.**

1. THE FERTILIZED EGG AND ZYGOTE STAGE

The eggs were fertilized just prior to the start of the laboratory period. The cytoplasmic mass of the egg is approximately 80 μm in diameter (if *S. purpuratus* is used; *Lytechinus pictus* eggs are 100 μm). Directly against the cell membrane you will see a clear, gelatinous layer about 3-4 μm in diameter. It is the **hyaline layer** and, as you will later observe, it acts during cleavage as a solid wall which encloses all the blastomeres resulting from the cleavage divisions. The blastomeres are firmly attached to the hyaline layer. Exterior to the hyaline layer is a space known as the **perivitelline space** (Figure 3.1). Beyond the perivitelline space is the **fertilization envelope**, which provides a tough sheath that protects the embryo until it hatches. The hyaline layer and fertilization envelope are not present in the unfertilized egg; they arise as a result of the cortical reaction triggered by the sperm.

To watch the zygote divide, adjust your light source by partially closing the iris diaphragm and adjusting the condenser height so that the hyaline layer and fertilization envelope are clearly visible. *S. purpuratus* zygotes divide approximately 90 minutes after fertilization; *Lytechinus pictus* zygotes proceed much faster, and divide after about 60 minutes. You may see the dumbbell-shaped mitotic apparatus in the center of the egg. It will appear clearer or lighter brown than the rest of the cytoplasm. The mitotic apparatus is the transitory structure that engages the chromosomes and pulls them to the opposite ends of the cell as mitosis proceeds. By focusing through the egg you may be able to resolve spindle fibers. If you see an egg beginning to elongate, i.e., lose its spherical shape, that is the signal that you are about to witness **cytokinesis** (cytoplasmic division), i.e., division of the two cell stage.

2. CLEAVAGE STAGES

The newly fertilized egg is a large single cell with a single diploid nucleus. This single-celled zygote must transform into a multicellular individual with each cell retaining its diploid condition. The embryo achieves the multicellular condition through a process called **cleavage**. Cleavage involves a series of mitotic cell divisions without cell enlargement during interphase. Thus, while the genetic material replicates with each division, there is no increase in the amount of cytoplasm. Consequently, each daughter cell is about half the size of the cell from which it came. After about six to eight division cycles the cells are much nearer the size of somatic cells found in the mature organism. Because there has been no cell growth, the multicellular young embryo is approximately the same size as the zygote.

The Two- and Four-Cell Stage. If two- or four-cell embryos are available, note that as the embryo divides the hyaline layer does not enter the division furrow but remains perfectly spherical (Figures 3.2, 3.3). Note also that the first and second divisions are vertical, i.e., they divide the embryo from animal to vegetal pole, thus producing blastomeres of equal size and shape.

Eight-, 16-, and 32- cell stage. The third cleavage division is horizontal or equatorial, i.e., it occurs at right angles to the first and second so that after the third division, the embryo consists of eight equal blastomeres. The fourth division is extremely interesting, and you should attempt to observe it. The fourth division occurs in the same plane as the third. At one pole, four of the blastomeres of the eight cell stage divide unequally, giving rise to four **micromeres** and four **macromeres**. The other four blastomeres of the eight cell stage divide equally, giving rise to eight **mesomeres**.

The 16-cell embryo thus consists of eight mesomeres, which are on top of four macromeres, which are on top of four micromeres. The micromeres will give rise to the primary mesenchyme, which will form the embryonic skeleton. At the fifth cleavage division, the number of meso-, macro-, and micromeres doubles and the embryo appears as a solid mass of blastomeres, called the morula (Figure 3.4).

Find a living 16-cell or 32-cell embryo.

3. THE BLASTULA STAGE

With additional cleavage divisions, a fluid-filled cavity arises at the center of the morula. (In echinoids, as in many animal groups, the morula stage is very short, or even non-existent, because the beginnings of a fluid-filled blastocoel appear even while cleavage is in process. The morula stage is probably most distinct in mammals.) The cavity expands as cell division continues until the embryo reaches a stage called a **blastula** (Figure 3.5), in which there is only a single layer of cells surrounding the fluid-filled cavity or **blastocoel**. In advanced blastulae you will note that the blastomeres are larger at one pole. The thickened area is the **vegetal pole**; the opposite, thinner wall is the **animal pole**. Each blastomere grows a cilium on its external surface. The cilia begin to beat in synchrony and the blastula spins inside the fertilization envelope. At hatching, the blastomeres secrete a proteolytic hatching enzyme that digests the fertil-

ization envelope. The blastula then swims out of the hull of the fertilization envelope into the surrounding sea water.

Shortly before or after hatching you will see a clump of cells forming at the vegetal pole. This is the primary mesenchyme cell mass which characterizes the so-called **mesenchyme blastula stage** (Figure 3.6). Draw a mesenchyme blastula in your notebook.

The mesenchyme mass enlarges inwardly and cells at the top of the mass migrate from the mass up the wall of the blastocoel where they form a band of cells about one-third the distance from the vegetal to animal pole. These are the **primary mesenchyme cells** which secrete the embryonic skeleton, and are derived from the micromeres. The starfish embryo does not show an ingression of primary mesenchyme cells, since there is no primary mesenchyme comparable to that found in the sea urchin.

4. THE GASTRULA STAGE: FORMATION OF THE ARCHENTERON

The process of gastrulation (Figures 3.7, 3.9) begins when the layer of cells at the vegetal pole bends inward, forming an invagination on the outer wall of the vegetal pole. The depression in the vegetal wall is the **blastopore**, which later becomes the anus of the larva. The invaginating epithelium elongates to become tubular and extends toward the animal pole. This tube is the **archenteron** or primitive gut tube. It extends by the process of convergent extension, wherein the forceful intercalation of cells results in the elongation of the epithelium. The tip of the archenteron finally reaches the animal pole where it connects to the **stomodeum** or larval mouth opening. Later, the primordia of the mesodermal coelomic sacs appear as two pouches that bulge from the tip of the archenteron. These sacs will give rise to the secondary mesoderm, which forms all mesodermal structures in the adult. The appearance of these pouches is more easily observed in the starfish embryo. Make a drawing of a mid-gastrula in your notebook.

The embryo has lost its rounded shape and now appears prismatic (prism stage). The **gastrula**

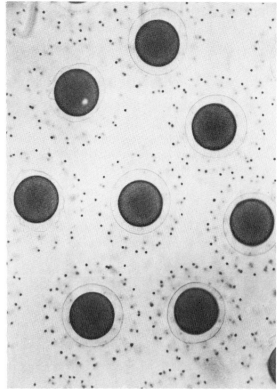

Figure 3.1. *Zygote of the west coast sand dollar,* Dendraster excentricus. The small cells embedded in the egg jelly around each egg are pigment cells. (Fig. 3.1 to 3.8 coutesty of V.D. Vacquier).

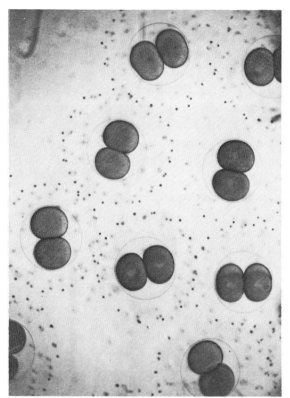

Figure 3.2. *Two-cell stage of* D. excentricus.

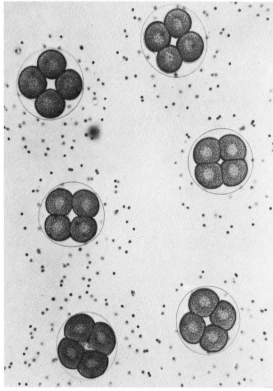

Figure 3.3. *Four-cell stage of* D. excentricus.

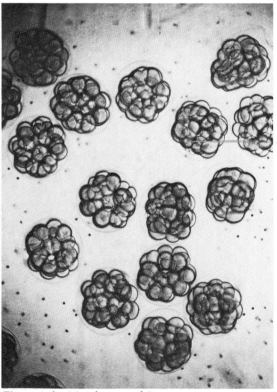

Figure 3.4. *Morula stage of* D. excentricus.

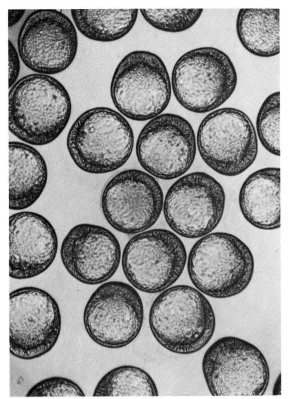

Figure 3.5. *Blastula stage of* D. excentricus.

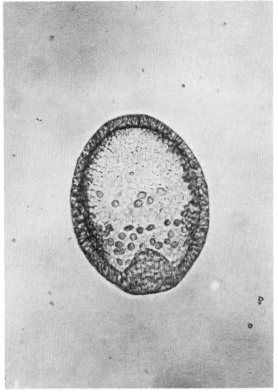

Figure 3.6 *Mesenchyme blastula of* D. excentricus.

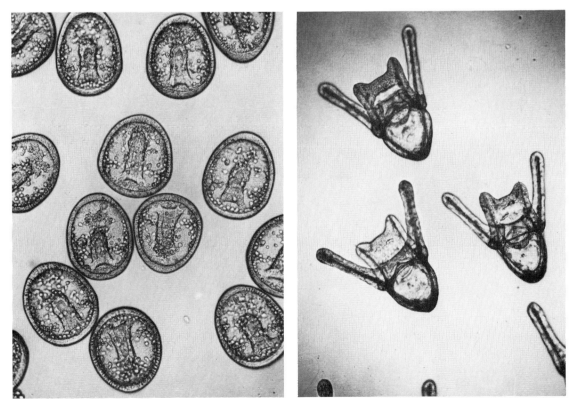

Figure 3.7. *Gastrula stage of* D. excentricus.　　**Figure 3.8.** *Pluteus larvae of* D. excentricus.

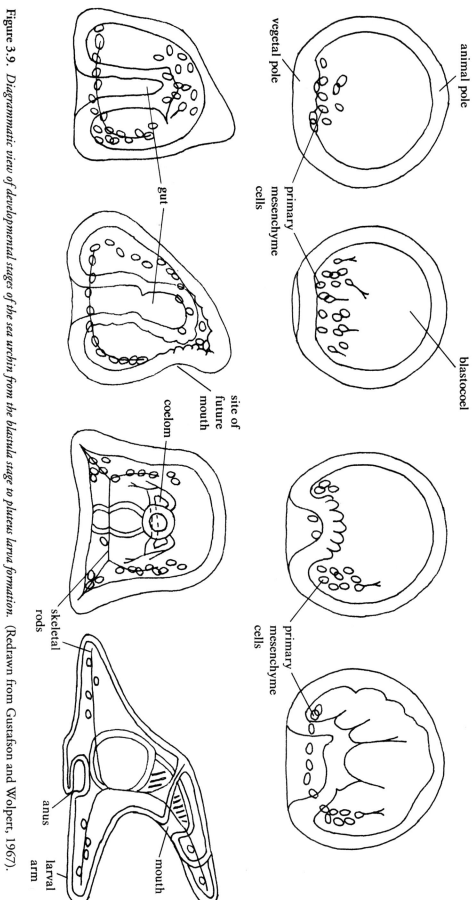

Figure 3.9. *Diagrammatic view of developmental stages of the sea urchin from the blastula stage to pluteus larva formation. (Redrawn from Gustafson and Wolpert, 1967).*

animal pole

vegetal pole

primary
mesenchyme
cells

blastocoel

primary
mesenchyme
cells

gut

site of
future
mouth

coelom

skeletal
rods

anus

larval
arm

mouth

has a completed digestive tube which will differentiate into a functional gut, permitting the pluteus larva to begin to feed on phytoplankton. By this time the skeletal rods are well developed. [The late gastrula in the starfish is pear-shaped, rather than prismatic, and lacks the skeletal rods.] Draw a prism stage embryo.

5. THE PLUTEUS

The pluteus (Figure 3.8) is the pelagic (free-swimming) larva of the sea urchin. In order to observe them better, you can slow down their movement by adding glycerol to the embryo suspension on your slide. Your TA will give you detailed instructions. Observe the arms of the pluteus stage. Pigmented cells are numerous and are arranged in patterns. The skeletal rods are extremely strong; for their dimensions, they are thought to be among the strongest natural structures known. The gut has differentiated into a three-part digestive system. From mouth to anus it consists of a muscular **esophagus** (pear-shaped), a round, thin-walled **stomach**, and a small, bulbous **intestine** connecting with the anus (Figures 3.8, 3.9). The esophagus exhibits rhythmic contractions. Ciliary currents direct phytoplankton into the esophagus where digestion begins. The pluteus feeds for several months and undergoes additional changes in form. It finally settles and metamorphoses to the form of the adult sea urchin. In sea urchins, the early morphogenetic events are concerned only with the development and establishment of the larval body. The body of the adult develops completely de novo in the body cavity of the pluteus. Unlike the situation in vertebrates, none of the larval structures of the sea urchin is incorporated into the body of the adult. [The stage of the starfish comparable to the pluteus is **a bipinnaria larva**. The larva is oval, lacks skeletal rods, and when mature possesses a series of lobes or arms. As with the pluteus of the sea urchin, the bipinnaria swims by means of epidermal cilia.]

It is interesting to note that the larval stages of echinoderms are bilaterally symmetrical, yet the adult shows pentameral (five-fold) radial symmetry. The adult radial symmetry is not a regression to a more primitive form, but an adaptation to sessile life. Make a drawing of a pluteus larva.

III. Amphibian Embryos

Amphibian embryos at various stages of development will be available for study. Proper illumination of your stereo microscope is required to see embryonic structures in any detail. Focus the light source on the embryo so that illumination is as strong as possible. In some cases structural detail is seen to best advantage with oblique illumination. Experiment with illumination conditions to obtain those most suitable for the material.

A. Major Features of Early Amphibian Development

1. NEWLY FERTILIZED EGGS

Prior to fertilization, the mature frog egg has completed only the first of two meiotic divisions necessary to reduce the chromosome number to the haploid state, and is arrested in metaphase of the second meiotic division. The region where the chromosomes are positioned appears as a round, whitish spot at the animal pole. The pigment granules in this area of the egg's cortex have been displaced by the spindle apparatus that has assembled for the completion of meiosis. Shortly after fertilization, the second meiotic division occurs, producing a haploid female pronucleus that then fuses with the male pronucleus contributed by the sperm. The two meiotic divisions result in the production of four nuclei, but only one of these ends up in the egg. The others are sequestered in tiny structures called **polar bodies**. With strong oblique illumination, the polar bodies can be seen in fertilized frog eggs as small spheres sitting atop the egg near the animal pole.

The **gray crescent** appears in *Rana pipiens* embryos about one or two hours after fertilization, and is visible even after two or three cleavages have occurred. The gray crescent can be seen as a grayish streak on one side of the egg between the pigmented animal hemisphere and the whitish vegetal region. Its color is almost the intermediate of the colors of the animal and vegetal poles. The gray crescent marks the site

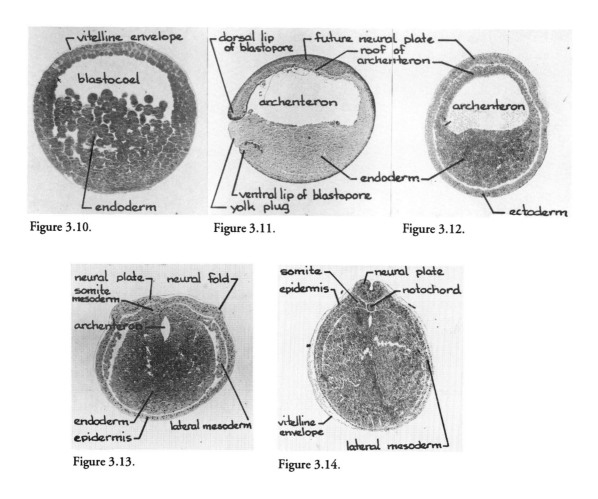

Figure 3.10. **Figure 3.11.** **Figure 3.12.**

Figure 3.13. **Figure 3.14.**

Figure 3.10. *Sagittal section through a blastula-stage newt embryo.*

Figure 3.11. *Mid-sagittal section through a mid-to-late gastrula-stage frog embryo.* This embryo is at approximately the same stage of development as that shown in Figure 3.15.

Figure 3.12. *Transverse section through a gastrula-stage newt embryo.* The mesoderm has not separated from the endoderm in this embryo. At this stage of development, the embryo is arranged as two layers of tissue: The surface layer is the ectoderm, which completely surrounds the embryo. The inner layer consists of future mesoderm (roof of the archenteron), and the future endoderm (floor of the archenteron). At a later stage, mesoderm and endoderm will separate from each other along the ventro-lateral margins of the archenteron. Mesoderm will migrate ventrally between ectoderm and endoderm; endoderm will migrate dorsally under the mesodermal roof of the archenteron. The net result of these late movements of gastrulation is the establishment of a three-layered arrangement of the embryonic tissues.

Figure 3.13. *Neural plate-stage newt embryo.* The dorsal migration of endoderm is complete, resulting in a gut cavity enclosed entirely by endoderm. The ventral migration of mesoderm has begun, but the expanding mesoderm has not yet reached the ventral midline of the embryo. Neural plate and neural folds are evident in the dorsal ectoderm. This embryo is approximately at the same stage of development as that pictured in Figure 3.16.

Figure 3.14. *Neural fold-stage newt embryo.* The mesoderm has subdivided into notochord, somites, and lateral mesoderm. The neural folds are closing over the dorsal midline. This embryo is approximately at the same stage of development as that pictured in Figure 3.17.

where the blastopore will later form. In *Xenopus* embryos, the gray crescent is difficult to observe until cleavage has begun.

2. CLEAVAGE STAGE

From the supply of embryos provided, try to find some very early cleavage stages, i.e., 2- or 4-cell stages. The first cleavage furrow initially appears in the animal hemisphere, then extends down through the vegetal hemisphere, dividing the egg into two **blastomeres**. If you can find one of these early stages, be sure you watch it long enough to see one of the cleavage divisions occur; it may take 10-15 minutes of careful watching to see it, but it will be well worth the wait! The second cleavage, like the first one, is in a vertical plane; the third cleavage is horizontal; the fourth is vertical again. Try to estimate how many cleavages have occurred in the youngest embryo you can find. Were all the cleavages in your embryo synchronous? How can you tell? Draw a cleavage stage embryo.

3. BLASTULA STAGE

After a few cleavages have occurred, a cavity (called the **blastocoel**) develops in the interior of the embryo. This cavity subsequently enlarges to occupy most of the animal hemisphere.

Using a razor blade, carefully bisect a blastula stage embryo. Note the difference in size between the cells in the animal and vegetal hemispheres. Why is there a size difference? What are the other prominent differences between the cells in these two regions?

Compare your dissected material with a sectioned preparation (Figure 3.10)

4. GASTRULA STAGES

Gastrulation begins with the movement of a small numbers of cells from the surface into the interior of the embryo. The first indication that this process is underway is the appearance of a crescent-shaped ridge that runs approximately parallel to the equator of the embryo and lies between the equator and the vegetal pole. This ridge is the **dorsal lip of the blastopore**. It is at this site that migration of surface cells into the interior occurs by the processes of invagination and involution. The cells on the surface of the animal hemisphere move as a continuous sheet toward the dorsal lip of the blastopore, where they roll into the interior of the embryo. A new cavity, the **archenteron**, is formed by the involution of the surface material; this cavity displaces the blastocoel. The dorsal lip gradually advances over the surface of the vegetal hemisphere, and its ends elongate to form the lateral lips. This extension of the ends of the blastopore continues until the blastopore becomes circular (see Figure 3.15). A part of the vegetal cytoplasm protrudes from the blastopore for a time; this is called the **yolk plug**. The blastopore marks the future posterior end of the animal; the anus will later form near this site. As a result of the gastrulation process the three germ layers, **ectoderm**, **mesoderm** and **endoderm**, are formed.

Dissect carefully a gastrula-stage embryo with a razor blade. Compare your dissected material with a sagittal section of a gastrula-stage embryo, identifying the blastocoel, archenteron, future notochord, and future neural plate. Use Figures 3.11, 3.12, and 3.15 for reference. Draw one of your sagittal sections of a gastrula.

5. NEURULA STAGES

Beginning with the process of neurulation, the major features of the vertebrate body plan, which you observed in a previous period, are starting to be laid out. Externally, the first sign of neurulation is the formation of a flattened area on the surface of the late gastrula (see Figures 3.13, 3.16, 3.17). This area is called the **neural plate** (also called the medullary plate), and is produced by a thickening in the ectoderm in this region. The plate is bounded on its periphery by elevations called **neural folds**. Later, the plate will roll up and the neural folds will fuse in the dorsal midline to form the **neural tube** (future brain and spinal cord) (Figures 3.13, 3.14, 3.17, 3.18). Although this process is called **neurulation**, in reference to the formation of the neural tube from the ectoderm, important changes in the mesoderm and endoderm are also taking place. The notochord is being blocked out from the chordamesoderm, the **lateral mesoderm** is moving ventrally, and the endoderm is fusing along the dorsal midline to form a new roof of the archenteron.

31

Once again, using a razor blade, bisect a neurula-stage embryo with a transverse section. Identify the notochord, the archenteron, the endoderm, the mesoderm, and the future somites. Compare what you observe with prepared slides of cross sections of neurula embryos. Make a drawing of your transverse section.

Once the neural closure stage is reached, additional structures are discernible. On each side, there are transverse grooves at the site of the future gills; these are the first and fifth visceral grooves. Two swellings on the ventral surface of the anterior end mark the developing mucous glands. The mouth will later form between them. Identify the somites along the dorsal surface.

B. Fertilization of *Xenopus* Eggs

1. *Xenopus* females will be injected prior to the lab with 800 I.U. of human chorionic gonadotropin to induce ovulation. You will be provided with approximately 20-30 eggs in a dry 60-mm pertri dish. Inspect the eggs through a dissecting scope. Remove unhealthy eggs (broken eggs, eggs with uneven pigmentation, eggs without a small white spot at animal pole, etc.) with forceps. If this is too difficult, due to the egg jelly, wait until the eggs are fertilized and dejellied, and then remove with a pipet. To insure proper development, the final number of eggs should not exceed 30 per petri dish.

2. For maximum fertilizations, be sure that the dry eggs are in a monolayer on the petri dish. Your TA will gently apply enough sperm suspension to just cover the eggs (the entire bottom of the dish should not be filled with sperm). Take up any sperm that runs off the eggs with a pipet and place back onto the eggs. Remember to note the time that the sperm was added. This is the insemination time. All times listed below are relative to this time (i.e., 10 minutes P.I. is 10 minutes postinsemination).

3. At about 3 minutes P.I., fill the petri dish with enough 20% Modified Ringers solution (MR) to cover all eggs and observe the eggs under a dissecting scope for the next 10-15 minutes. Usually 95% of the eggs will undergo fer-

tilization. In eggs that have been fertilized, the pigmented animal hemisphere will contract towards the animal pole and will appear darker in color. This is due to the **cortical contraction** during egg activation. Another feature of the egg activation is **fertilization envelope liftoff** (the envelope can be visualized after dejellying the eggs). Also, notice a small, darkly pigmented spot in the animal hemisphere which looks like a dent or pucker in the egg surface. This is the **sperm entry point** and denotes the future ventral surface of the embryo. The opposite side, where gastrulation is initiated, marks the future dorsal surface of the embryo. Egg activation is usually complete within 20 minutes P.I.

C. Dejellying Zygotes
Dejellying zygotes is done to provide better visualization, and to permit experimental manipulation of the embryos.

Step 1. Approximately 20 minutes P.I., fill the petri dish containing the zygotes with enough 2% cysteine in F1 (pH 7.8) to cover the eggs and cover the dish. Working over a sink, swirl the petri dish gently but continuously until all the zygotes no longer stick to the bottom of the dish (3-5 minutes).

Step 2. Tilt the dish slightly so that all the zygotes roll to the edge of the dish, then pour out the cysteine solution (without pouring out any of the zygotes). Alternatively, you may use a pipet to remove the solution.

Step 3. Add fresh cysteine solution and swirl the dish again until all zygotes are dejellied (approximately 2-3 minutes). Check the progress of dejellying frequently by tilting the dish to one side so that the zygotes roll to the edge. Completely dejellied zygotes will touch one another with no visible space between them.

Step 4. Pour or pipet off the cysteine solution and gently add 20% MR solution to the dish. Swirl briefly, pour or pipet off, and repeat six times (a total of seven rinses in 20% MR). Be careful not to pour out any of the zygotes! When done, fill the dish with 20% MR and cover with a clean lid.

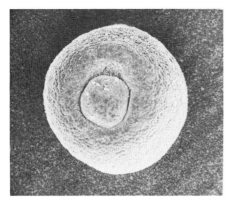

Figure 3.15.

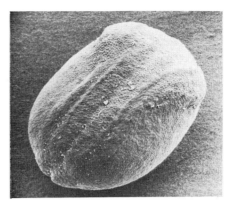

Figure 3.16.

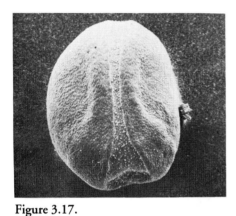

Figure 3.17.

Figure 3.18.

Figure 3.15. *Scanning electron micrograph of the ventral surface of a mid-gastrula-stage* Xenopus *embryo.* The yolk-plug is surrounded by the dorsal and lateral lips of the blastopore. The dorsal lip is uppermost in this figure.

Figure 3.16. *Scanning electron micrograph of the dorsal surface of a neural plate-stage* Xenopus *embryo.* The anterior end of the embryo (marked by the expanded portion of the neural plate) is directed toward the upper right corner of the figure. The posterior end (marked by the remainder of the blastoporal groove) is situated at the lower left corner.

Figure 3.17. *Scanning electron micrograph of the dorsal surface of a neural fold-stage* Xenopus *embryo.* The anterior end is facing the lower margin of the figure.

Figure 3.18. *Scanning electron micrograph of the anterior surface of a neural closure-stage* Xenopus *embryo.* The neural folds have met and are fusing over the mid-brain and hind-brain regions but have not closed completely in the forebrain region.

The complete dejellying process should take no longer than <u>10 minutes including rinsing</u>. A long exposure to cysteine will damage the zygotes. If it takes much longer, either you are not swirling vigorously enough, or the pH of the cysteine is not correct.

D. Care and Feeding of Tadpoles

The embryos generated from our observations of fertilization will be used in a subsequent study of metamorphosis. Below are the steps needed to care for them.

1. From 0-5 days old
   -change 20% MR daily
   -remove any dead embryos
   -keep total number of embryos per dish to
     20 or less.

2. 5+ days old
   -transfer tadpoles to large, clear containers
     with air stones for O_2
   -use tap water instead of 20% MR
   -change water twice a week or whenever
     cloudy
   -feed ground Frog Chow or fresh spinach
     every day (feed only as much as they will
     clean up in 24 hours)
   -remove any tadpoles that die
   -females will be much larger than males

Development of the Chick Embryo: A General Study

I. Introduction

The pattern of early morphogenesis exhibited by higher vertebrates[1] (reptiles, birds, and mammals) is quite different from the amphibians, which we have already studied, owing to differences in the content of yolk in the egg. One significant developmental difference as a result of yolk content, is the pattern of cleavage. Amphibian eggs contain a moderate amount of yolk, and during cleavage this yolk is parcelled primarily into the endodermal cells, although ectodermal and mesodermal cells also contain considerable amounts of yolk. But in the eggs of reptiles and birds, almost all the yolk is stored extracellularly, in a structure called the **yolk sac**. The cleavage furrow cannot cut through the entire ovum, but instead extends only through a small disc of yolk-free cytoplasm sitting on top of the large sphere of yolk. In spite of the very small amount of yolk found in mammalian eggs, the embryos develop as though a large amount of yolk is present, as indeed it was in their immediate evolutionary ancestors.

In addition to modifying the patterns of cleavage, the presence of a bulky yolk mass necessitates numerous morphological adaptations in the later development of the embryo. Not only must the embryo develop systems for mobilizing and absorbing the yolk, it must also modify its body structure and morphogenetic movements to accommodate the yolk mass.

The chick embryo provides an excellent model for embryological studies, because it portrays developmental themes common to amniotes. Chicken eggs are inexpensive and readily available, and the embryos are quite hardy. We will, therefore, spend a good deal of time observing these embryos, learning their basic anatomy, and studying various features of their development by experimentation.

II. Fertilization and Cleavage

The yellow part of the chicken egg, commonly called the "yolk," is really a very large single cell prior to first cleavage. This cell is surrounded by a plasma membrane similar to that found on the boundary of all cells. The plasma membrane is overlaid by a tough non-cellular **vitelline envelope**. The nucleus of the egg is encased in a tiny circular patch of cytoplasm sitting on top of the yolk mass. This cytoplasm, termed the **blastodisc**, is only 3 mm in diameter,

---

1. Because the embryos in these classes possess a set of extraembryonic membranes, including an amnion, they are often grouped together as amniotes. Fish and amphibians develop in water, lack these membranes, and are called anamniotes.

but it is destined to develop into both the embryo and the extraembryonic membranes.

Fertilization in chickens is internal,[2] and cleavage begins while the egg is still in the oviduct. As the egg passes down the oviduct it is wrapped in the watery albumin (the egg white) and encased in a calcareous shell.[3] As cleavage proceeds, the blastodisc becomes subdivided into smaller and smaller cells. Cells in the central region of the blastodisc become separated from the underlying yolk by a space called the **subgerminal cavity**. The portion of the blastodisc lying above the subgerminal cavity is called the **area pellucida** because of its translucent appearance in whole mount preparations (to be studied later). The periphery of the blastodisc in contact with the yolk is called the **area opaca**. Once it has become multicellular, the blastodisc is called a **blastoderm**.

The blastula stage in birds, as in other vertebrates, is completed with the formation of the **blastocoel** cavity. The avian blastocoel forms by a process called **delamination**. During delamination, some cells separate from the undersurface of the blastoderm and sink into the subgerminal cavity to form a loose sheet of cells called the **hypoblast**.[4] The remaining upper part of the blastoderm forms a compact epithelium called the **epiblast**. The space between the hypoblast and epiblast is the blastocoel; cells that make up the hypoblast correspond to the cells lining the floor of the blastocoel in amphibians.

III. Gastrulation in Birds

The result of gastrulation in all vertebrates is the transformation of the embryo into a three-layered arrangement of germ layers. As has already been seen in the amphibian embryo, this transformation is accomplished by the mass movements of cells from the surface of the blastula into the interior. Despite the rigidity imposed by the bulky yolk, gastrulation in birds also involves the movement of cells from the surface (epiblast) into the interior (blastocoel) of the embryo.

The location of the future germ layers in birds and the movement of this germ layer material during gastrulation can be related by constructing a fate map of the surface (i.e., the epiblast) of the late blastula stage embryo. The fate map of the epiblast differs when different methods are used to construct the map. Some maps have been constructed by marking the epiblast with vital dyes or with tiny carbon particles that adhere to the surface of epiblast cells. These techniques have been less than satisfactory, however, because the dyes tend to diffuse away from their original site and the carbon particles can become dislodged from the surface of the labeled cells as gastrulation proceeds. Tritiated thymidine (^{3}H-TdR) has been used to mark or label the cells prior to gastrulation. In this procedure, a fragment of the epiblast is cut from a donor embryo that has been exposed to ^{3}H-TdR. The radioactive fragment is then transplanted to a site on the epiblast of a nonradioactive host embryo that corresponds exactly to the site on the donor from which the fragment was taken. The fate of the cells in the radioactive fragment can then be reliably followed through the process of development by autoradiography.

The fate map derived from ^{3}H-TdR labeling has demonstrated that the relative positions of the future avian organs and tissues are in the same relative positions as in amphibians. For example, the future notochord maps adjacent to the future endodermal material that will line the gut. Likewise, the future notochord is contiguous with the future neural ectoderm which, in turn, is contiguous with the future epidermal ectoderm, etc.

---

2. A single mating is sufficient to fertilize many eggs, even though ovulation and laying occur only about once a day. The hen's reproductive tract has a number of sperm-storage compartments called sperm glands. The sperm are collected at these sites, maintained in a viable condition, then released at the time of ovulation.

3. For an interesting and well-written account of shell formation, see "How an Eggshell is Made," by T. G. Taylor in Scientific American, March, 1970.

4. Note that the process of blastocoel formation in birds resembles the process in amphibians and all other eggs with smaller amounts of yolk. In all cases, the cells that are to line the blastocoel must become nonadhesive over a part of their surface from their neighbors and thereby allow a fluid-filled cavity to form.

The onset of gastrulation is signaled by a thickening of one edge of the rim of the blastoderm. This thickening, called the embryonic shield, marks the future posterior end of the embryo and is the region where presumptive mesoderm is found. This thickened region of the blastoderm begins to elongate and rapidly forms the **primitive streak.** (Figure 4.2). A variety of labeling techniques show that all cells of the epiblast move toward the primitive streak during gastrulation. Once they reach the primitive streak, the epiblast cells move into the blastocoel cavity. These cells are destined to give rise to the endoderm and mesoderm. The primitive streak is therefore analogous to the blastopore of amphibians, although no opening ever forms. Because there is no open blastopore in birds, the movement of cells into the blastocoel cavity during gastrulation is a different process and is called **immigration.** Immigration appears to be a modification of the simpler invagination process seen in embryos that develop from eggs with little yolk like the sea urchin.

The first cells to move into the blastocoel cavity are the future endoderm cells that will form the

lining of the gut. Once inside the blastocoel i.e., beneath the surface of the epiblast, these cells invade and mingle with the central region of the hypoblast. They will ultimately give rise to the embryonic endoderm. The rest of the hypoblast forms part of the extraembryonic endoderm of the yolk sac. The future mesodermal cells also immigrate to the interior of the blastocoel when they reach the primitive streak. Once in the blastocoel, these mesodermal cells spread laterally as two wing-like sheets that intrude between the epiblast and the hypoblast (Figure 4.1). Immigration ceases when all presumptive mesoderm material has moved to the interior, leaving only ectoderm on the surface.

At the anterior end of the primitive streak there is a thicker and slightly elevated region known as **Hensen's node**, sometimes called the primitive knot or primitive node (Figures 4.2 through 4.5). Hensen's node is homologous to the dorsal lip of the amphibian blastopore because the presumptive notochord immigrates over it. The lateral edges of the streak correspond to the lateral lips of the amphibian blastopore, because the presumptive somites and lateral plate meso-

Figure 4.1. *Pattern of Gastrulation Movements.* The solid arrows represent epiblast cells moving into the embryo. Dotted arrows indicate the movement of the internalized cells. "A" indicates the position of the future oral plate. (Redrawn from Tuchmann-Dutlessis.)

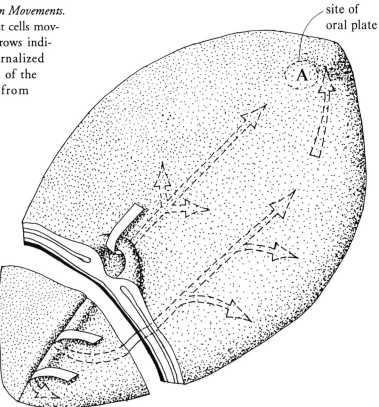

site of
oral plate

derm move over the lateral edges of the streak on their way to the interior. As gastrulation proceeds the streak begins to regress or shorten, with Hensen's node shifting toward the posterior end of the embryo. As the node retreats, the notochord is laid down in its wake (Figures 4.4 to 4.10). In some descriptions the newly formed notochord is misleadingly called the head process.

The process of gastrulation, i.e., the inward movement of material from the epiblast, continues as the primitive streak regresses toward the posterior end of the embryo. Gastrulation is not fully completed in the chick until about 20-21 somites have been formed, i.e., until about 50 hours of incubation, at which time the primitive streak finally disappears (Figure 4.24).

As gastrulation continues in the posterior region of the embryo, the three germ layers that have already been established in the anterior end of the embryo begin to form the primordia of the major organ systems. The first such development occurs in the ectoderm. A few hours after the anterior end of the notochord is established, the ectoderm lying above it thickens to become the neural plate. The neural plate continues to develop in an anterior-to-posterior direction, becoming progressively longer as the underlying notochord elongates in the wake of the receding primitive streak (Figures 4.6, 4.8, and 4.10). At its periphery, the neural plate elevates and forms **neural folds** (Figures 4.6 and 4.7). The neural folds subsequently fuse along the dorsal midline and form the **neural tube** (future **spinal cord**) (Figure 4.10). The fusion process also proceeds in an anterior-to-posterior sequence, as illustrated by the 26-hour embryo shown in Figure 4.9. At this stage, the neural folds in the head region have completely fused to form a tube which is already shaping into the brain. In the midregion of the same 26-hour embryo, fusion of neural folds is still in progress, while in the tail region the neural folds have not yet formed.

Immediately prior to the formation of the neural plate in the anterior region of the embryo, the head begins to take shape. Head formation begins with a crescent-shaped fold in the ectodermal region of the epiblast, just anterior to the notochordal process. The formation of this fold, called the head fold, intitiates the transformation of the flat embryo into a tubular structure (Figures 4.5-4.7, and 4.10). The conversion of the previously flat embryo into a tube is much like what happens when you attempt to push your finger through a flat sheet of rubber: the rubber conforms to the contours of your finger to form a tube, but remains flat in all other area. Now examine the large-scale model of the 24-hour embryo in the laboratory to appreciate the three-dimensional arrangement of these changes.

As the ectodermal head fold continues to develop, the underlying flat sheet of endoderm is also converted into a tube (Figure 4.8). The zone of transition between the endodermal tube (or foregut) and the more posterior region of the endoderm that has not yet formed a tube is called the **anterior intestinal portal** (Figures 4.7 through 4.12). The foregut lengthens and its outline becomes more apparent as the head continues to be shaped by the rearward movement of the head fold (Figures 4.10 and 4.12).

Other primordia of the embryonic body are rapidly delineated during this period. For example, the mesoderm lying on either side of the notochord becomes separated into blocks of tissue called **somites** (Figures 4.6 to 4.10). Also by 28 hours of development the heart and major regions of the primitive brain can be identified (Figure 4.10).

Figures 4.2 through 4.10 have all been printed at the same magnification. Compare these two figures to gain an understanding of the relative growth of various regions of the embryo and to appreciate the enlargement of the embryo as a whole. The fate of all the components observed in this early period will be examined in detail in the subsequent sections of this chapter.

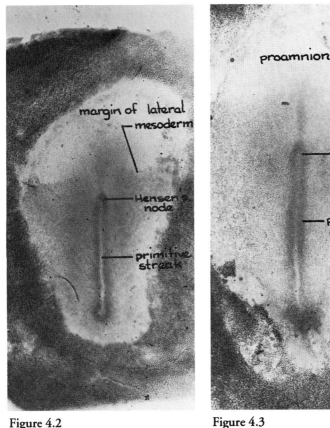

Figure 4.2

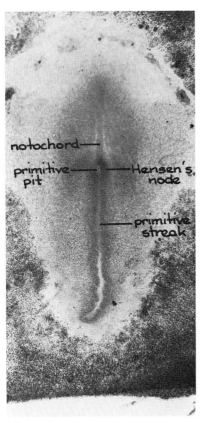

Figure 4.3

Figure 4.2 and 4.3. *Primitive Streak Stage (13-15 hours).* Figures 4.2 through 4.14 are photomicrographs of the whole mounts, viewed from above. Keep this perspective in mind as you study them. The primitive streak appears as two parallel condensations of tissue separated by a lighter area. The long axis of the primitive streak is parallel to the long axis of the area pellucida. The condensations of tissue forming the lateral margins of the primitive steak are called the primitive folds and are homologous to the lateral lips of the amphibian blastopore. The light central region is called the primitive groove and is homologous to the amphibian blastopore.

Figure 4.4.

Figure 4.4. *Notochordal Process Stage (18 hours).* The primitive streak is longer than at 13-15 hours. Anterior to the primitive streak is a condensation of tissue beneath the epiblast known as the notochordal process. This is the anterior end of the notochord. The remainder of the notochord has yet to immigrate over Hensen's node.

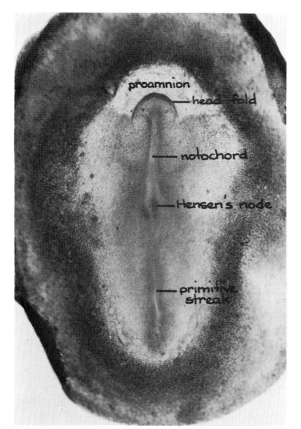

Figure 4.5

Labels in figure: proamnion, head fold, notochord, Hensen's node, primitive streak

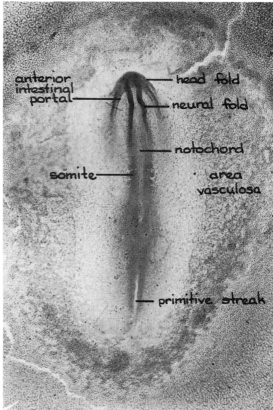

Figure 4.6

Labels in figure: anterior intestinal portal, head fold, neural fold, somite, notochord, area vasculosa, primitive streak

Figure 4.5. *Head Fold Stage (20 hours).* The primitive streak has begun to shorten. The notochordal process is longer and more condensed than at 18 hours. Hensen's node is visible as a condensed area surrounding a central depression (the primitive pit). The anterior tip of the embryo is lifted off the surface of the epiblast by a crescent-shaped fold called the head fold. The anterior limit of the mesoderm that is spreading between the epiblast and hypoblast is about at the level of the head fold. The clear, mesoderm-free area called the proamnion is visible anterior to the head fold.

Figure 4.6 and 4.7. *Early Somite Stage (23 hours).* The head fold has lifted further off the blastoderm. The posterior margin of the head fold is delimited by the anterior intestinal portal, which appears as a crescent-shaped line. The cavity within the head fold and anterior to the anterior intestinal portal is the primordium for the foregut. The anterior regions of the neural plate have formed by this stage. Arching up from the borders of the neural plate are the ridge-like neural folds. The folds give the appearance of a dark, hairpin-like structure lying above the head fold. The notochord is visible as a long, narrow structure lying beneath the neural plate. Posterior to the neural plate, the mesoderm on either side of the notochord has condensed into somites. Further posteriorly, the primitive streak can be seen. Blood-forming regions are beginning to appear in the peripheral region of the mesoderm called the area vasculosa.

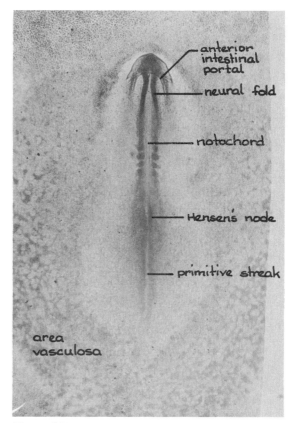

Figure 4.7

Labels in figure: anterior intestinal portal, neural fold, notochord, Hensen's node, primitive streak, area vasculosa

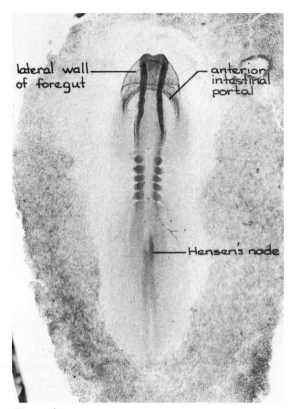

Figure 4.8

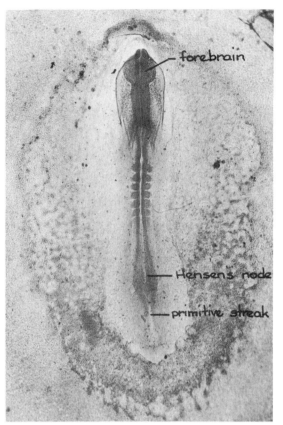

Figure 4.9

Figure 4.8. *Early Neural Closure Stage (26 hours).* The neural folds have closed in the anterior portion of the embryo, and regionalization of the brain has begun. The head fold has progressed more posteriorly. Additional somites are present. Note that the primitive streak is much shorter than before.

Figure 4.9. *Late Neural Closure Stage (28 hours).* The neural tube has closed, except for the posterior region of the spinal cord. The line of fusion of the neural tube (the neural suture) is open at the tip of the forebrain (the neuropore). The three major regions of the brain (prosencephalon, mesencephalon, and rhombencephalon) are clearly established. Hensen's node now lies close to the posterior end of the embryo; the primitive streak is barely visible. The heart has formed beneath the foregut, and the vitelline veins, which will soon transport blood from the extraembryonic circulation to the heart, lie just ahead of the anterior intestinal portal.

Figure 4.10. *Optic vesicle stage (32 hours).* The optic vesicles (optic cups) have developed from the prosencephalon. The vitelline veins are more clearly defined as they enter the heart region, and the heart is now clearly visible as a bulge on the right side. The primitive streak has reduced further.

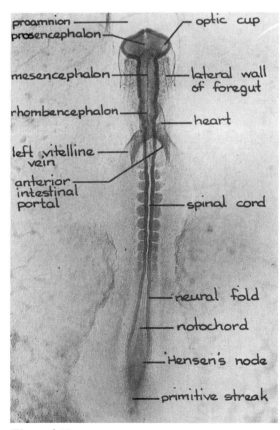

Figure 4.10

41

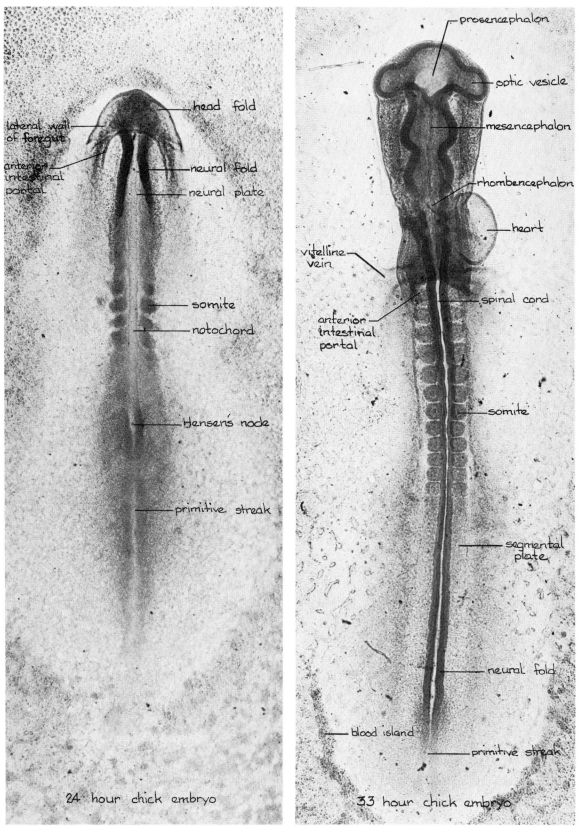

Labels on left figure (24-hour chick embryo):
- head fold
- lateral wall of foregut
- anterior intestinal portal
- neural fold
- neural plate
- somite
- notochord
- Hensen's node
- primitive streak

24 hour chick embryo

Labels on right figure (33-hour chick embryo):
- prosencephalon
- optic vesicle
- mesencephalon
- rhombencephalon
- heart
- vitelline vein
- spinal cord
- anterior intestinal portal
- somite
- segmental plate
- neural fold
- blood island
- primitive streak

33 hour chick embryo

Figure 4.11. *24-hour Chick Embryo.* Whole mount.　　**Figure 4.12.** *33-hour Chick Embryo.* Whole mount.

42

IV. Major Events in Morphogenesis: Survey of Whole Mounts and Cross Sections

The aim of this section is to give you a brief guided tour through young chick embryos at key stages of development. For the time being you should ignore the minute details of anatomy, and concentrate instead on learning the major landmarks and the morphogenetic events that transform the chick gastrula (primitive streak stage) into an embryo in which most of the major organ primordia are present (72 hours). Later, when you have the major events well in mind, we will return to examine the formation of some organs in greater detail. The major terms that you should know are in **boldface** type.

A. Prepared slides of 33-hour chick embryos

It is fundamentally important to understand the gross changes in body shape that occur during

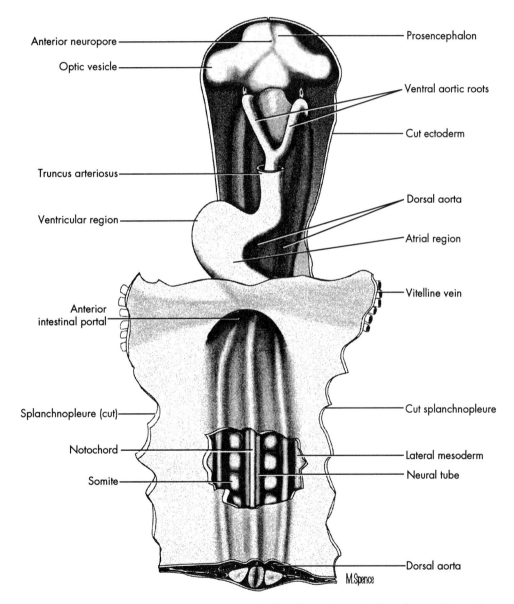

Figure 4.13. *Diagrammatic ventral view of a 33-hour chick embryo.* The splanchnopleure of the yolk sac anterior to the anterior intestinal portal, the ectoderm of the ventral surface of the head, and the mesoderm of the pericardial region, have been removed to show the underlying structures. (Adapted from *Early Embryology of the Chick*, 4th ed., by B. M. Patten.)

early development of the chick. Hence, we will first examine whole mount preparations of 33-hour chick embryos. A sound grasp of the three-dimensional arrangements of embryonic structures at this stage will make your subsequent study far easier. After you have finished the whole mount, the same stage will be re-examined in greater detail in serial cross sections.

NOTES ON PROCEDURES: Examine the whole mounts under the highest power of a binocular dissecting microscope. Illuminate the embryo from below if possible. For greatest comfort, use the frosted side of the substage mirror. Adjust the mirror and light so that the amount of light coming through each eyepiece of the microscope is the same. Some structures are most easily seen from the underside of the embryo or when the whole mount preparation is tilted at an angle. If a dissecting microscope is used, the slide can easily be inverted or tilted on the microscope stage for optimal viewing. If a dissecting microscope is unavailable, a compound microscope will be satisfactory, but be sure to use ONLY the lowest power objective. There is little advantage to viewing the underside of embryos with a compound microscope, and <u>UNDER NO CIRCUMSTANCES SHOULD THE SLIDE BE TILTED.</u>

Many important details of developmental anatomy cannot be seen in whole mounts and it becomes necessary to examine serial cross sections of the embryo. In the slides you will study, the entire embryo has been sectioned from the anterior to posterior direction and these sections have been mounted on slides in rows from left to right. There are usually several slides in the set for any one stage; hence the most anterior section is Slide #1, Row #1, Section #1 (upper left-hand corner). The first section on Slide #2 begins where the last section of Slide #1 left off, etc.

Each set is composed of cross sections from a single embryo. Thus it is extremely important that you do not mix slides from your set with your neighbor's set. To help you keep your slides in order, each set has an identification number on the label.

The most serious difficulty students have with studies of cross sections is perceiving how the two-dimensional sections can provide an understanding of the original three-dimensional embryo. This difficulty is most easily overcome if frequent and careful comparisons are made between the sections and the whole mounts and large-scale models available in the laboratory.

1. 33-HOUR WHOLE MOUNT

Examine the 33-hour whole mount under the highest power of a binocular dissecting microscope. If one is not available, a compound microscope will be satisfactory, but be sure to use <u>only</u> the <u>lowest</u> power objective. Refer to the diagrams and photomicrographs included in this chapter for aid in locating structures.

The anterior end of the embryo juts forward above a clear region of the blastoderm, the proamnion (Figure 4.5). This finger-like projection over the proamnion is called the **head fold** and is the beginning of the head. The dominant feature within the head fold is the **neural tube**, which is beginning to differentiate distinct subdivisions or **vesicles** (see also Figures 4.12 and 4.23). The region of the neural tube at the tip of the head is the forebrain or **prosencephalon**, characterized by the **optic vesicles** bulging laterally. Just posterior is the midbrain or **mesencephalon**. The hindbrain or **rhombencephalon** is the most posterior vesicle. Its walls display a series of constrictions (**neuromeres**) characteristic of this region of the brain in most vertebrates. The rhombencephalon extends posteriorly to the level of the 5th somite; the remainder of the neural tube is the **spinal cord**. Note that in its most posterior region the neural "tube" is not yet closed but is instead a trough or **neural groove**; this feature will be more apparent in cross sections. Just posterior to the neural groove the remnants of the **primitive streak** can be seen. The **notochord** runs anteriorly from the streak beneath the floor of the neural groove. Trace the notochord toward the head. It usually appears in the anterior region of the embryo as a thin dark line lying in the midline beneath the brain. How far forward does it extend?

The **somites** appear in the middle third of the embryo as blocks of darkly staining material on either side of the neural tube. How many are there? Somites are blocked out in an anterior-posterior direction; the newest ones are therefore the most posterior. Can you see regions in which new somites are being formed? The region of mesoderm from which the somites are formed is called the **segmental plate**. At this stage only the somites in the head and neck have been formed; those of the trunk and tail will develop later from the segmental plate. The first five somites, lying lateral to the rhomben-cephalon, will contribute primarily to the muscle, bone and connective tissue structures of the neck. The remaining somites will give rise to the vertebrae and other bony structures of the back, all the striated muscle, and connective tissue structures of the back. By now you have probably noticed that the anterior end of the embryo is more advanced in development than the more posterior end. This pattern will continue during the time the main organ primordia are being laid down. Thus, early development proceeds in an **antero-posterior sequence**. As you have already observed, the anterior neural tube has already differentiated into the three main brain vesicles, while posteriorly you find a neural groove that has not yet closed to form a neural tube.

Return now to the head fold region. Note that as this finger-like projection juts forward, it lifts off the yolk, and is encased by ectoderm on all boundaries except posteriorly (Figure 4.11). This will become more evident in your study of the large-scale models of the whole mounts and in the cross-sections. You will find it useful to refer to these models frequently during your study of the chick embryo. They will be particularly helpful in providing a better understanding of the three-dimensional arrangement of structures within the embryo.

The loosely packed head **mesenchyme** in the head fold has a stippled appearance and lies between the neural tube and the outer body wall ectoderm. By looking carefully you should be able to see the faint outline of the tubular **foregut** lying beneath the mesencephalon and rhombencephalon. The anterior end of the foregut is obscured by the brain, but, in good preparations, the lateral walls can be seen to run parallel to the lateral body walls of the head fold. The foregut terminates posteriorly at the **anterior intestinal portal**, the location of which can be identified by a crescent-shaped line in front of (or just below) the most anterior somite. This can be seen best if the slide is inverted (see also Figure 4.13). (Take care not to crack the coverslip.) The primordia for the midgut and hindgut lie posterior to the anterior intestinal portal. In the 33-hour embryo, these are still flattened epithelial sheets lying over the yolk.

The **heart** can be seen bulging to the right of the rhombencephalon. Once circulation begins, blood will be returned to the heart from the extraembryonic circulation by the two large **vitelline veins** lying just anterior to the anterior intestinal portal (Figure 4.13). Invert the embryo again. Can you see where these two veins fuse into a single tube in the midline?

The relatively clear halo around the entire embryo is the **area pellucida**; it appears this way because this area of the blastoderm is usually yolk-free when removed from the egg. Peripheral to this clear area is the mottled **area opaca**.[5] The stained blotches in the area opaca are the **blood islands** in which the formation of the first blood cells is underway.

Draw the 33-hour chick whole mount in you notebook. Identify and label all the structures printed in bold face type. Use the models in the laboratory to help you visualize the 3-dimensional structure of the embryo.

2. 33-HOUR CROSS SECTIONS
a. *Quick survey*
Quickly scan the entire series of sections from anterior to posterior under the dissecting microscope. Keep the 33-hour whole mount available to use for reference. The first few sections on your slide will probably consist only of extraembryonic tissue. The first part of the embryonic body that comes into view is the anterior tip of the head fold which appears as a solid mass of cells lying above the extraembryonic blastoderm.

---

5. That part of the area opaca containing the blood islands and the developing extraembryonic blood vessels is called the area vasculosa. The most peripheral part of the area opaca is not vascularized and is called the area vitellina: the latter can only rarely be seen in mounted preparations.

One or two sections further back the elliptically shaped, hollow structure is the **prosencephalon**. A few sections later, the **optic vesicles** are seen as lateral diverticula of the prosencephalon. In sections immediately posterior to the optic vesicles, the anterior end of the **foregut** appears as a cavity lying beneath the brain. Note that the epithelium that lines the foregut is only about half as thick as the epithelium lining the prosencephalon.

As you pass more posteriorly through the embryo, the foregut widens laterally to become the **pharynx**. The lopsided structure lying beneath the pharynx is the **heart**, which will be studied in detail later. Toward the posterior end of the pharynx the head ectoderm becomes continuous with the extraembryonic ectoderm. As you continue moving posteriorly, notice that the floor of the tubular foregut disappears and the gut spreads like a flat sheet on the surface of the yolk. The point at which the gut makes the transition from tube to flattened sheet is the anterior intestinal portal. The flattened gut lying posterior to the **anterior intestinal portal** is the **midgut**. All along the midgut region the **spinal cord** is obvious as the dorsal thick-walled tube surrounding an elliptical cavity. Directly beneath the spinal cord, the **notochord** is seen as a small rounded structure. **Somites** are also apparent as large blocks of mesoderm lying on either side of the spinal cord. The somites connect to more lateral mesodermal tissue that spreads to the extraembryonic blastoderm. As more posterior sections are examined, the neural tube can be seen to open dorsally. This is the neural groove. In this region the neural folds have not yet met and fused to form the neural tube. Continue scanning the sections posteriorly to the point where the groove, lateral mesoderm and notochord all appear to fuse. This is the region of the **primitive streak**. In this region, mesoderm is still invaginating from the surface of the blastoderm.

b. Examination with compound microscope
Now re-examine the 33-hour cross sections under the **compound microscope**. Set up the microscope correctly for Köhler illumination.

Be certain the top lens of the condenser is swung out of the light path for the scanning objective but is in position for all higher power objectives. For your study of the cross sections, use as a guide the photomicrographs of representative cross sections.

In the anterior-most region of the embryo, note that only ectoderm and endoderm are present in the extraembryonic blastoderm lying immediately below the head fold; this is the **proamnion**[6] region. You can see mesoderm between the ectoderm and the endoderm on either side of the proamnion area. Notice the globular spongy yolk that is attached to the endoderm. You should be able to identify the neural tube and notochord. The crescent shaped opening ventral to the neural tube is the **pharynx**. Identify the mesenchyme between the neural tube and the surface ectoderm.

Locate a section through the heart region of the embryo, similar to the one shown in Figure 4.16. Find all the labeled structures on your slide (you may have to examine several sections). Determine how blood from the heart reaches the dorsal aorta by tracing the heart tube anteriorly (i.e., move from right to left on your slide). The lumen of the heart tube becomes markedly smaller as the vessel continues beneath the pharynx as the **ventral aorta** (Figure 4.15). The ventral aorta bifurcates near the region where the head begins to lift off the blastoderm. Near this point of bifurcation, notice the oral plate, a thickening in the floor of the pharynx that touches the ectoderm on the ventral surface of the head (Figures 4.1 and 4.14). This structure will perforate later in development to form the anterior opening to the **gut tube**. Move further anteriorly and find the **first aortic arches**, the blood vessels connecting the ventral and dorsal aorta. Because their walls are extremely thin, blood vessels appear as spaces in the mesenchyme. Notice the appearance of the **optic vesicles** of the prosencephalon.

Return now to a section similar to that pictured in Figure 4.16. Follow the heart tube posteriorly, observing that it splits into the two **vitelline**

---

6 . This is another one of embryology's misleading terms. Actually, the proamnion has nothing to do with the amnion, which will form later. The proamnion is simply an area in the extraembryonic region which has not yet been invaded by mesoderm. The lack of mesoderm accounts for the clear appearance of this area in whole mounts.

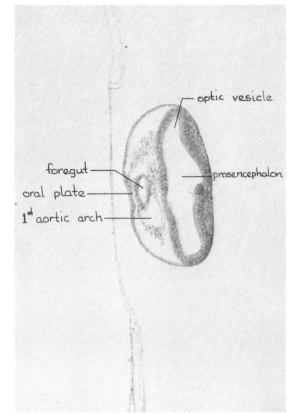

Figure 4.14

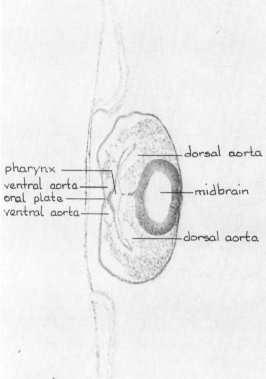

Figure 4.15

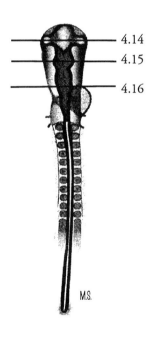

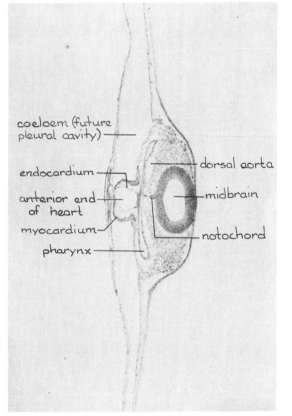

Figure 4.16

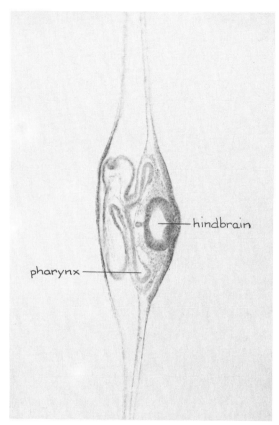

Figure 4.17

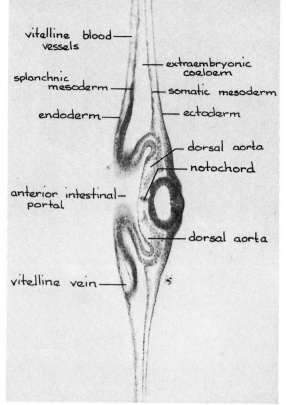

Figure 4.18

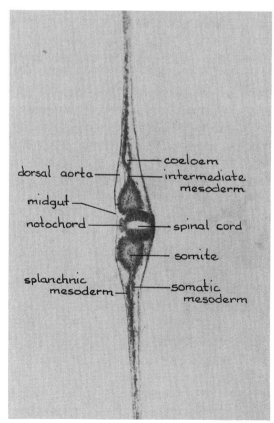

Figure 4.19

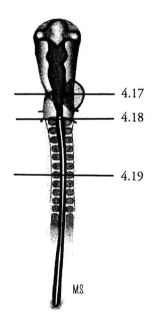

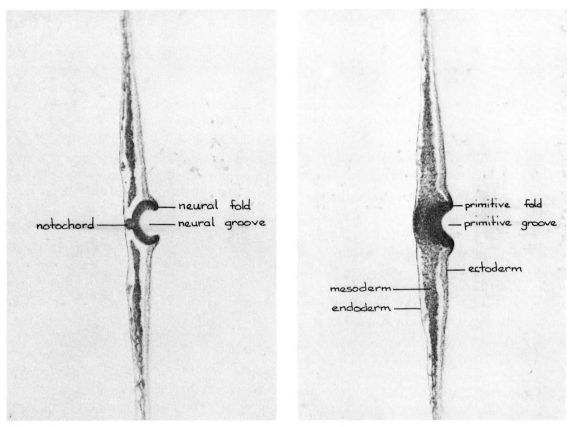

Figure 4.20

Figure 4.21

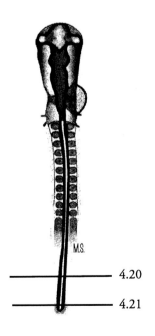

veins. What happens to these veins in more posterior sections? Note that the vitelline veins fuse just anterior to the **anterior intestinal portal,** the point at which the gut becomes tubular (Figure 4.18). A few sections posterior to the portal the **somites** can be seen to be connected to a double layered sheet of lateral mesoderm lying between the ectoderm and endoderm of the blastoderm. The dorsal layer of mesoderm that is opposed to the ectoderm is called **somatic mesoderm,**[7] while that associated with endoderm is termed **splanchnic mesoderm** (Figure 4.18). Which of these mesodermal layers contains blood vessels? The space between these two layers is the **coelom,** or body cavity (Figure 4.19). Note that it is a cavity totally lined by mesoderm. The somites and the lateral mesoderm are connected by a relatively thin bridge of **intermediate mesoderm,** which will later form structures of the excretory system.

c. Map embryo on grid

Using the ruled form provided, you will construct a "grid-map" of the 33-hour chick embryo from analysis of cross sections. This exercise is intended to help you become familiar with the three dimensional relationships of the different organ systems in the developing embryo. In addition, it will help you to learn those parts of the embryo that can be used as "landmarks" owing to their spatial relationship with other structures which are more difficult to identify with certainty. You will also begin to familiarize yourself with the germ layer derivation of the embryonic structures you map by drawing the structures on your map in colors that are codes for particular germ layers. Use yellow for endodermal structures, red for mesoderm and blue for ectoderm; if you wish you may use green for neural ectodermal structures.

First, look at your slides with the un-aided eye, count the number of sections in each row and draw a dashed vertical line along the top horizontal axis of your grid to mark the end of each row. Each centimeter square will represent 10 sections on your slides; section numbers are indicated across the top. Now, notice that the

left-hand side of the grid is labeled for the three germ layers, and on the right side of the grid, several lines or spaces are labeled for specific structures derived from the three germ layers. Please use this arrangement to map specific structures in your embryo, as will be described below. The sample grid-map (Figure 4.22) will give you an idea of how it will look.

Orient your slides so that the dorsal surface of the embryo is at the top of the viewing field (yolk sac down). Start at the first (most anterior) section of your embryo. Mark on your map the level at which the head first appears. Notice that the embryo is lying above the extraembryonic blastoderm, which consists of an upper ectoderm layer and a lower endoderm layer. The extraembryonic layers underlying the head comprise the proamnion. The space between the proamnion and the head is the subcephalic pocket. Mark the extent of this pocket on your map. Notice that the extraembryonic tissue on either side of the proamnion has four cell layers: the dorsal two are the somatopleure (upper layer is ectoderm, lower is somatic mesoderm) the ventral two cell layers make up the splanchnopleure (upper layer is splanchnic mesoderm, lower endoderm). The space between splanchnic and somatic mesoderm is a true coelom (completely lined with mesoderm).

Continue to follow your sections in an anterior to posterior sequence, identifying all the structures in your embryo that are printed in bold face type in the text, and extending the colored lines on your grid map for the ectoderm-, mesoderm-, and endoderm-derived structures. Label individual landmarks like optic vesicles, 1st somite, etc., by widening the line if the structure is a bulge in the germlayer, or alternatively by placing dots above or below the germlayer line to delineate the extent of those structures. You may find it helpful to follow one organ system at a time completely through the embryo from head to tail rather than trying to draw in all the systems at once.

---

7. The layer of surface ectoderm and the somatic mesoderm that together make up the body wall is called somatopleure. The layer of splanchnic mesoderm and endoderm that together will contribute to development of most digestive organs is called splanchnopleure. At this stage both somatopleure and splanchnopleure extend outside the embryo and form the extraembryonic membranes. For more detailed discussion, see page 62.

 50

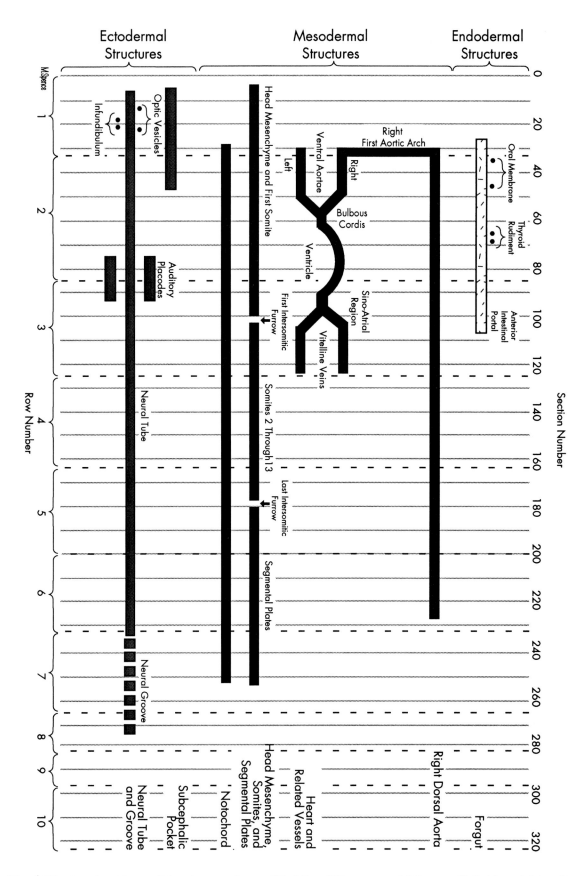

Fig. 4.22 *Sample grid map.* (Mapping exercise modified from Watterson and Schoenwolf, *Guide and Atlas of Vertebrate Embryology*, Fifth Edition, Burgess Publishing Company.)

When you are done you should have a "grid-map" which shows the major endodermal (gut), mesodermal (heart and blood vessels, head mesenchyme, somites and segmental plates) and ectodermal (subcephalic pocket and neural tube and groove) structures and their relative positions in your embryo. Notice that posterior to the neural groove lies Hensen's node and the primitive streak (Figure 4.21). In this area the embryo is still undergoing gastrulation, and the three germ layers are not yet differentiated, illustrating the anterior-posterior developmental gradient in the embryo.

Turn in your whole mount drawing and grid-map at the end of lab today even if they are not complete. They will be checked and returned next lab. Mistakes noted on the drawings must be corrected before the notebook is turned in for grading at the end of the quarter.

B. Prepared slides of 48-hour chick embryo
1. WHOLE MOUNTS
Two dramatic changes have occurred since the 33-hour stage: (1) the anterior half of the embryo has turned over on its left side and is lying on the yolk, and (2) the tip of the head has bent posteriorly to lie near the heart (Figure 4.24). (You could think of it as the embryo

tucking his chin!). As a result of this bending or **flexure** of the head, the ectoderm on the ventral surface of the head is thrown into a fold or pocket that comes to lie near the tip of the foregut. The ectodermal pocket will form the lining of the mouth cavity or **stomodeum** of the animal.

The brain has become further subdivided into five regions (Figure 4.23). The prosencephalon is now divided into two vesicles: the **telencephalon** and the **diencephalon**. The telencephalon lies at the very tip of the head and, due to head flexure, almost touches the heart. It is separated by a constriction from the diencephalon, which is the posterior part of the prosencephalon. The eyes (**optic cups**) are now present as outpocketings of the diencephalon, and are diagnostic of this region of the brain. The **mesencephalon** remains undivided and, due to head flexure, lies at the most anterior end of the embryo. Posterior to the mesencephalon, the two subdivisions of the hindbrain can be seen. The **metencephalon** lies just posterior to the constriction separating the midbrain and hindbrain. Its roof or dorsal surface is stained as heavily as its walls. Next comes the **myelencephalon**, the most posterior region of the hindbrain. Two features mark the myelencephalon:

PRIMARY BRAIN
VESICLES

SECONDARY BRAIN
VESICLES

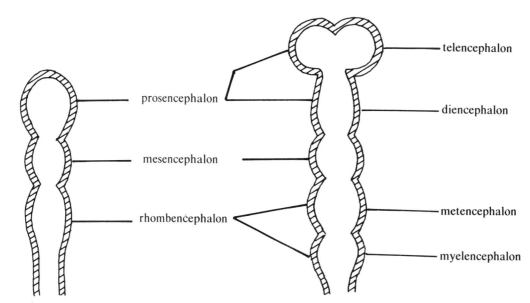

Figure 4.23 *Diagrammatic View of the Brain Vesicles in the 33-hour and 48-hour Chick Embryo.* (Redrawn from Moore).

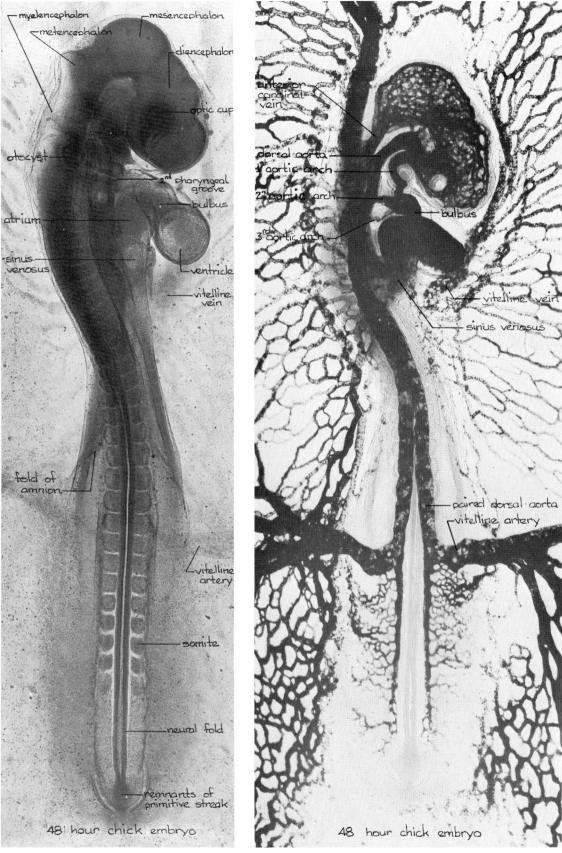

Figure 4.24 labels:
myelencephalon
metencephalon
mesencephalon
diencephalon
optic cup
otocyst
2nd pharyngeal groove
bulbus
atrium
sinus venosus
ventricle
vitelline vein
fold of amnion
vitelline artery
somite
neural fold
remnants of primitive streak

48 hour chick embryo

Figure 4.25 labels:
anterior cardinal vein
dorsal aorta
1st aortic arch
2nd aortic arch
bulbus
3rd aortic arch
vitelline vein
sinus venosus
paired dorsal aorta
vitelline artery

48 hour chick embryo

Figure 4.24. *48-hour Chick Embryo.* Whole mount.

Figure 4.25. *48-hour Chick Embryo.* Injecting dye into the circulatory system highlights the pattern of circulation.

53

Table 4.1. Some Landmarks and Derivatives to Help You Find Your Way Around Your Embryo's Brain.

		Landmarks	Adult Derivatives
PROSENCEPHALON	Telencephalon	nasal placodes lateral ventricles	olfactory epithelium lateral ventricles cerebral hemispheres
	Diencephalon	optic cups optic stalks infundibulum	retina optic nerves posterior (neural) lobe of the pituitary
		third ventricle epiphysis	third ventricle pineal body (gland) thalamic region of the brain
MESENCEPHALON	Mesencephalon	aqueduct of Sylvius	aqueduct of Sylvius optic tectum midbrain
RHOMBENCEPHALON	Metencephalon	fourth ventricle thick roof	fourth ventricle cerebellum pons
	Myelencephalon	fourth ventricle neuromeres thin roof	fourth ventricle medulla oblongata choroid plexus

(1) it has a very thin (hence, lightly stained) transparent roof which may be folded inward in some embryos and (2) its walls retain the series of constrictions (**neuromeres**) seen earlier. Usually these can best be seen on the underside of the embryo. Lying lateral to the myelencephalon are the otocysts or future inner ears. A summary of development and fate of brain components is found in Table 4.1. For additional information see also the chapter on Development of the Nervous System.

The **pharyngeal grooves** and **pharyngeal pouches** lie just ventral to the otocysts. The grooves are slit-shaped shallow depressions in the surface ectoderm: usually three can be seen on each side of the embryo at this stage. Each groove lies lateral to a **pharyngeal pouch**, an outpocketing of the endodermal lining of the pharynx. The ectoderm of the groove has pushed in slightly to meet the bulging wall of the pharyngeal pouch (endoderm). At the point of contact, the ectoderm and endoderm fuse to form a thin, transparent membrane. In the chick, there are ultimately four pairs of pharyngeal pouches and grooves. In all except the fourth, the thin membrane ruptures transiently, thus forming for a short time an opening between the pharynx and the exterior. This opening is termed the **pharyngeal cleft**[8] and corresponds to the gill cleft of aquatic vertebrates. The regions between pharyngeal clefts are called **pharyngeal arches.** The pharyngeal arches are the mesenchymal masses lying between the body wall ectoderm and the pharyngeal endoderm (foregut) in the embryo. Study this region in the large-scale models of the 33- and 48-hour chick to aid you in visualizing the three-dimensional arrangement of these structures. The pharyngeal region and its deriv-

8. The term pharyngeal is often used interchangeably with the terms visceral or branchial. Thus, pharyngeal pouch = visceral pouch = branchial pouch.

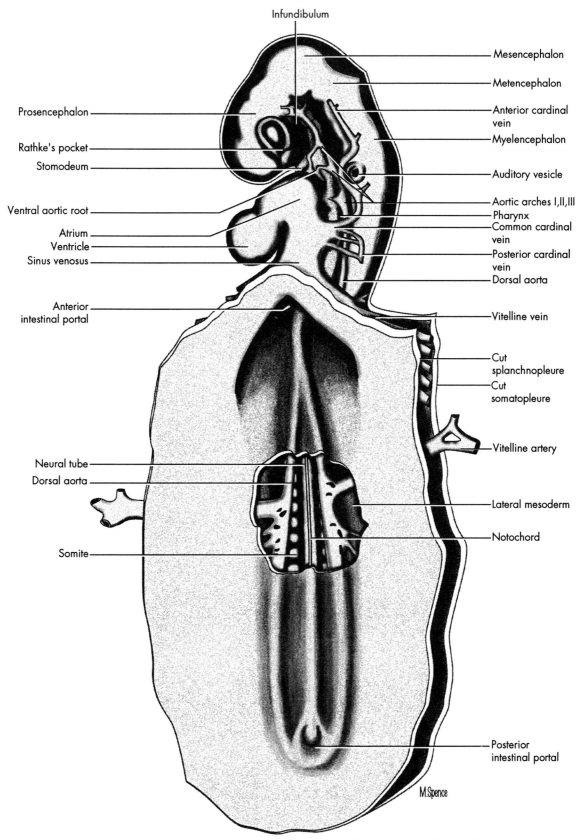

Figure 4.26 *Diagram of dissection of chick of about 48 hours.* The splanchnopleure of the yolk sac cephalic to the anterior intestinal portal, the ectoderm of the left side of the head, and the mesoderm in the pericardial region have been dissected away. A window has been cut in the splanchnopleure of the dorsal wall of the midgut to show the origin of the vitelline arteries. (Redrawn from *Patten's Foundations of Embryology*, Fifth Edition, B. M. Carlson, McGraw-Hill Book Company.)

55

atives will be considered in greater detail later in this chapter as you study the cross sections.

The endodermal **foregut**, of which the pharynx is a part, extends posteriorly to the **anterior intestinal portal**, which has become more V-shaped in outline since 33 hours. This can be seen best if the slide is turned over and held at an angle so that you can look down into the foregut through the anterior intestinal portal (Figure 4.26).

The heart, which begins development as a straight tube situated beneath the pharynx, aquires its final form by bending and folding. This folding was already apparent in the 33-hour embryo and is now more pronounced. Although the basic architecture of a simple tube is still obvious, the future subdivisions of the heart can now be identified. The most ventral portion is the **ventricle** which opens anteriorly into the **bulbus** (or **conus**) **arteriosus**. Blood flows from the bulbus into the **ventral aorta** and subsequently into the aortic arches lying within the visceral (pharyngeal) arches (Figures 4.25, 4.26). The aortic arch most readily apparent is the second, which lies between the first and second pharyngeal clefts, just above the bulbus. On the underside of the embryo, the first aortic arch can also be distinguished. Blood flows from the aortic arches to the dorsal aorta and then posteriorly in the embryo. The **vitelline arteries**, which carry blood from the dorsal aorta to the extraembryonic circulation, can be seen about two-thirds of the body length from the head.

Blood returns to the embryo from the extraembryonic circulation via the two **vitelline veins**. These are most visible near the point where they enter the heart. The vitelline veins enter the heart at the **sinus venosus**. The sinus venosus lies just anterior to the anterior intestinal portal and can be seen to best advantage if the embryo is examined from beneath. Blood flows from the sinus venosus to the atrium, then on to the ventricle and the circuit begins again. The path may be summarized as follows:

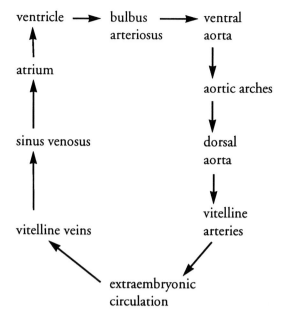

The pattern of blood flow you have just seen in the 48-hour whole mount is really the same as you observed in the 33-hour embryo. Compare the 33- and the 48-hour large-scale models to help you in understanding the change in form achieved by the folding of the cardiac tube. Also study the dye-injected embryo shown in Figure 4.25 and the line drawing in Figure 4.26.

Note the increased number of somites. Where is the somite that was nearest the segmental plate at 33-hours? Where is the segmental plate now?

Several other features of the 48-hour whole mount should be noted.
1) In the posterior regions, the body is beginning to lift off the yolk. As it does so, the ectoderm and mesoderm gradually fold in between the body proper and the underlying blastoderm; the folds are called the **lateral body folds** (Figure 4.27) and will participate in body wall formation.
2) In the most advanced specimens a tail fold has formed as the most posterior part of the embryo lifts off the yolk.
3) The hood-like **anterior amniotic fold**, a transparent membrane of ectoderm and mesoderm, is present. The posterior limit of this fold can usually be seen as a crescent-shaped boundary just dorsal to the embryo near the level of the anterior intestinal portal.[9] A posterior

56 9. In some specimens the posterior limit of the amniotic fold may be further advanced in its movement toward the caudal region of the embryo.

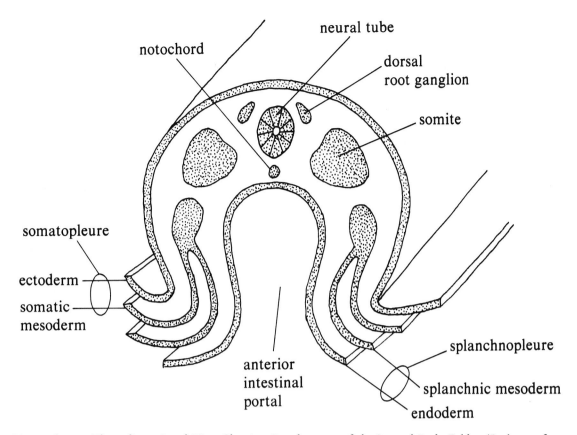

Figure 4.27. *Three-dimensional View Showing Development of the Lateral Body Folds.* (Redrawn from Tuchmann-Dutlessis.)

amniotic fold is later thrown up over the tail region where it meets and fuses with the anterior fold. More detail on the formation of the amnion may be found in the discussion of extraembryonic membranes beginning on page 62.

4) Blood vessels can be distinctly seen in the extraembryonic blastoderm.

By now you should have a clear three-dimensional picture of the gross morphological changes taking place. The large-scale models of the 33- and 48-hour whole mounts will be useful in clarifying any problems you may still have.

Now draw your 48-hour whole mount.

2. 48-HOUR CROSS SECTIONS

a. Quick survey

For a quick survey of the 48-hour cross sections, examine the slides under the dissecting microscope as was done with the 33-hour cross sections. Keep the 48-hour whole mount available and compare sections with the whole mount. As you scan your sections, look for diagnostic

structures to orient you in the embryo. Because the 48-hour embryo has turned, the sections are not exactly perpendicular. It is often helpful to make a rough sketch of the orientation your embryo had prior to sectioning. In most cases, the first sections go through the **mesencephalon**, which is the most forward portion of the embryo, due to the bending of the anterior regions of the head at the head (cephalic) flexure. As you move posteriorly in the embryo, the brain becomes hourglass shaped and divided into two separate vesicles. The anterior vesicle is the mesencephalon and the posterior vesicle (easily identified by its thin roof) is the **myelencephalon** (review Figure 4.23 and Table 4.1). Between the myelencephalon and mesencephalon is the rather poorly defined **metencephalon** (this is seen more easily in the whole mount). More anteriorly locate the **diencephalon** identifiable by its association with the optic cups. Anterior to the diencephalon is the **telencephalon**. This region of the brain is also more easily seen in the whole mount. The diencephalon and telencephalon have developed through subdivision of the forebrain (prosen-

57

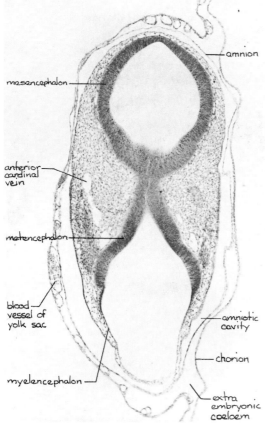

Figure 4.28

Labels for Figure 4.28:
- amnion
- mesencephalon
- anterior cardinal vein
- metencephalon
- blood vessel of yolk sac
- amniotic cavity
- chorion
- myelencephalon
- extra embryonic coeloem

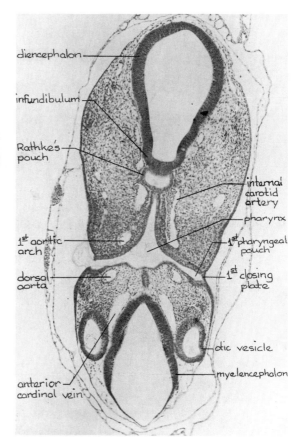

Figure 4.29

Labels for Figure 4.29:
- diencephalon
- infundibulum
- Rathke's pouch
- internal carotid artery
- pharynx
- 1st aortic arch
- 1st pharyngeal pouch
- dorsal aorta
- 1st closing plate
- otic vesicle
- anterior cardinal vein
- myelencephalon

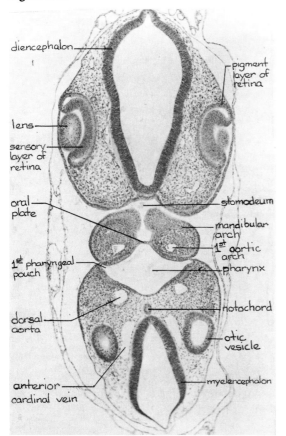

Figure 4.30

Labels for Figure 4.30:
- diencephalon
- pigment layer of retina
- lens
- sensory layer of retina
- oral plate
- stomodeum
- mandibular arch
- 1st aortic arch
- 1st pharyngeal pouch
- pharynx
- dorsal aorta
- notochord
- otic vesicle
- anterior cardinal vein
- myelencephalon

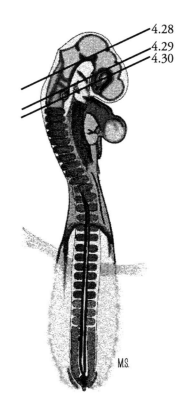

4.28
4.29
4.30

M.S.

58

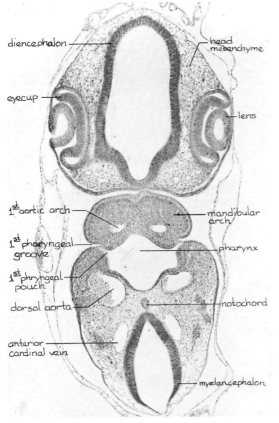

Figure 4.31

Labels for Figure 4.31:
diencephalon, head mesenchyme, eyecup, lens, 1st aortic arch, mandibular arch, 1st pharyngeal groove, pharynx, 1st pharyngeal pouch, dorsal aorta, notochord, anterior cardinal vein, myelencephalon

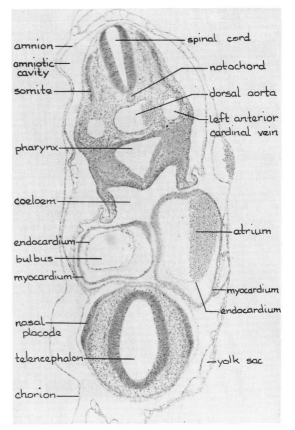

Figure 4.32

Labels for Figure 4.32:
amnion, spinal cord, amniotic cavity, notochord, somite, dorsal aorta, left anterior cardinal vein, pharynx, coeloem, endocardium, atrium, bulbus, myocardium, myocardium, endocardium, nasal placode, telencephalon, yolk sac, chorion

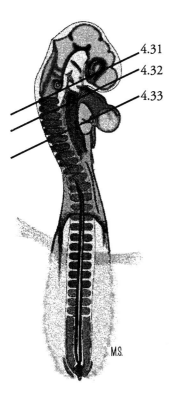

4.31
4.32
4.33

M.S.

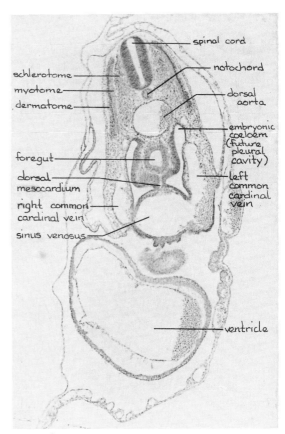

Figure 4.33

Labels for Figure 4.33:
spinal cord, schlerotome, notochord, myotome, dermatome, dorsal aorta, embryonic coeloem (future pleural cavity), foregut, dorsal mesocardium, left common cardinal vein, right common cardinal vein, sinus venosus, ventricle

59

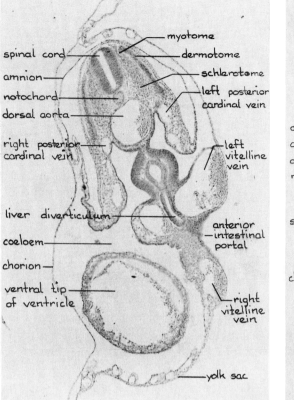

Figure 4.34

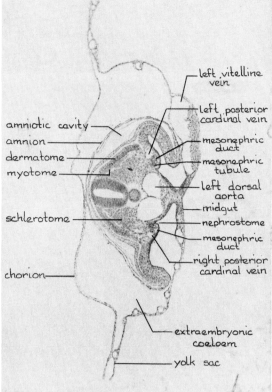

Figure 4.35

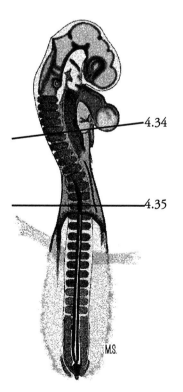

cephalon). [NOTE: If you are confused by the arrangement of the brain in these sections you should re-examine the whole mount and refer to Figure 4.23 and Table 4.1] Find the mesencephalon, myelencephalon and metencephalon. The appearance of myelencephalon and mesencephalon in the same section in the sectioned 48-hour embryo is a consequence of the head flexure that places the mesencephalon rather than the telencephalon at the most anterior end of the embryo and therefore the first part to be sectioned.

Examine the eyes in these sections. The **optic cup** and **lens** primordia are distinct. The lens vesicle forms by an inpocketing of the head ectoderm that lies directly above the optic vesicle. This invagination is especially apparent at this stage. The other most prominent sense organs in the 48-hour embryo are the **otic vesicles** (sometimes called **otocysts**). These are present as a pair of thick-walled sacs, situated on either side of the myelencephalon and open to the surface at their dorsal ends. Like the lens, each otic vesicle forms from a thickened region of head ectoderm that invaginates to form a cup-shaped structure. When the rims of the cup eventually fuse, the otic vesicles become hollow spheres that separate from the head ectoderm. The otic vesicles develop into the inner ears. Compare your sections to the 48-hour whole mount to appreciate how the eye cup and the otic vesicle can be present in the same cross section of the embryo.

Continue posterior to sections where the brain has divided into separate vesicles (Figure 4.28). Passing back from this point, the dorsal portion of the **pharynx** comes into view as a rounded space, surrounded by thick walls, lying between forebrain and hindbrain. Further caudad, the pharynx broadens out into the **first pharyngeal pouch**. The **first visceral arch** (mandibular)[10] lies anterior to the first pharyngeal pouch. A few sections caudad to the first appearance of the first pharyngeal pouch, the **mandibular arch**, which becomes the lower jaw, becomes separate from the head. The blood vessels passing up through the mandibular arches are the first **aortic arches**. Further posteriorly the pharynx narrows to become the posterior regions of the foregut (i.e., esophagus and stomach). The most posterior limit of the foregut is the **anterior intestinal portal**. Here the gut tube is not fused at its ventral aspect and now is open to the yolk. This flattened region of the gut is the **midgut**.

b. Examination with the compound microscope
Examine the slides a second time, this time with the compound microscope. Locate a section that matches as nearly as possible the one shown in Figure 4.30. Identify in your section all the structures labeled on the diagram including the **first visceral (mandibular) arch**, which will become the future lower jaw, the **first pharyngeal pouches** and **grooves**, the **pharynx** and **dorsal aorta**. Trace the pharynx posteriorly a few sections to find a point where its floor is only a thin membrane. This is the **oral plate**, a thin membranous partition which will rupture later in development, allowing the foregut to become continuous with the mouth. Note that the foregut extends a short distance anterior to the oral plate. The blood vessels on either side of this pre-oral region of the gut are the **first aortic arches** and at this point connect with the dorsal aortae. Follow the first aortic arches dorsally as they course through the **first visceral arches** (see Figures 4.29, 4.30, and 4.31). The **first pharyngeal pouches** (outpocketings of the pharynx) also appear in this region.

Move along posteriorly and find the point where the pharynx again bulges laterally to touch the surface ectoderm and form the **second pharyngeal pouches**. The third visceral arch is more difficult to distinguish at this stage and will be ignored for the time being.

Move forward a few sections to find the oral plate again. Now move posteriorly to trace the first aortic arches back to a point where they merge with a wide structure, the **bulbus arteriosus**, which hangs like a lopsided sac ventral to the pharynx. The connection between the bulbus and the ventricle can be seen a few sections further back. Pass now to the posterior end of

10. The first visceral arch splits into two by the ventral inpocketing of the stomaodaeum. The anterior portion gives rise to the upper jaw (maxilla) and is often called the maxillary arch. The posterior half forms the lower jaw (mandible) and is referred to as the mandibular arch.

the heart to find a section similar to the one in Figure 4.33.[11]

The two vessels seen emptying into the dorsal part of sinus venosus are the **common cardinal veins.** Each of these collects blood from an **anterior cardinal vein** running lateral to the brain. The anterior cardinals can be located most easily where they pass between the otic vesicles and the myelencephalon. Trace them back to where they join the common cardinals. Also emptying into each common cardinal vein is a **posterior cardinal vein**, which collects blood from body regions posterior to the heart. A few sections back from the point where the common cardinals enter the sinus venosus, the gut elongates ventrally to form the **liver diverticulum** (Figure 4.34). At the level of the liver diverticulum the sinus venosus bifurcates into two vessels, the **vitelline veins.** These carry blood from the yolk sac into the sinus venosus. Slightly farther back, the anterior intestinal portal appears. Trace the dorsal aorta posteriorly to the point where it bifurcates and then on to where the dorsal aortae become confluent with large vessels in the splanchnic mesoderm. These large vessels arising from the dorsal aortae are the **vitelline arteries.**

The **mesonephric ducts**[12] can be seen lying dorsal to the intermediate mesoderm (i.e., lateral to the somites). As in the frog, these ducts transport nitrogenous wastes removed from the blood by the kidney to the cloaca. At this stage the ducts are still growing toward the tail; their connection with the cloaca will be seen in the 72-hour chick. The kidney at this stage consists of a series of tubules, appearing as clumps or condensations of tissue, lying medial to the mesonephric duct. Find these tubules by tracing the ducts anteriorly. The most anterior tubules are pronephros-like in nature, so they also open into the coelom by way of a **nephrostome.** The posterior cardinal vein lies just dorsal to the mesonephric duct. The most posterior tubules are typical of the mesonephric type in that they have no connection with the coelom.

3. EXTRAEMBRYONIC MEMBRANES
Before proceeding further we should now give some attention to the set of extraembryonic membranes observed in chick embryos. The chorion and amnion are apparent at 48 hours; the beginnings of the allantois can usually be seen in 72-hour preparations. Be sure you can identify each of these membranes in your slides.

Amphibian and fish eggs, which are laid in water, have little difficulty in dealing with dessication, waste removal and gas exchange, because their aqueous environment provides a medium to facilitate these functions. Eggs of birds and reptiles, however, are laid on land, and have had to develop methods of coping with these problems. In addition to the development of a tough exterior shell to protect against drying and

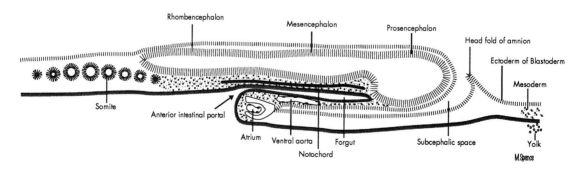

Figure 4.36. *Mid-sagittal Section of a 50-hour Chick Embryo Shows Head Fold of Amnion.* (Redrawn from *Patten's Foundations of Embryology*, Fifth Edition, B. M. Carlson, McGraw-Hill Book Company.)

11. The structure of the heart will be dealt with in more detail in the chapter on the circulatory system. The following discussion will only examine the major blood vessels.

12. At earlier stages these are called pronephric ducts because of their association with pronephric tubules. Mesonephric tubules are now forming and attaching to the pronephric duct. When this occurs the name is changed to mesonephric duct.

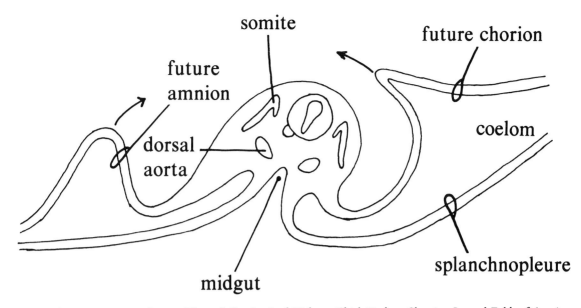

Figure 4.37. *Transverse Section Through Somite 5 of 52-hour Chick Embryo Showing Lateral Folds of Amnion* (Arrows).

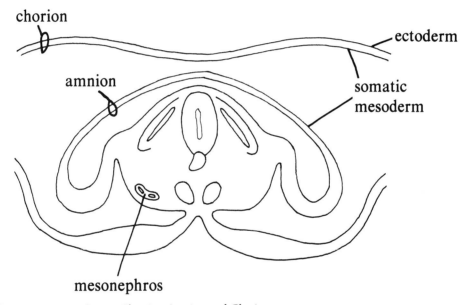

Figure 4.38. *Transverse Section Showing Amnion and Chorion.*

mechanical shock, avian eggs have developed a series of epithelial membranes to deal with the problems of dessication, respiration, and waste removal. The extraembryonic membranes of birds include the yolk sac and allantois (both of splanchnopleure origin) and the amnion and chorion (of somatopleure origin).

a. Formation of extraembryonic membranes in the chick
The amnion and chorion develop simultaneously by folding of the extraembryonic somato-

pleure. The somatopleure folds and arches over the head and dorsal midline to cover the forebrain by 40 hours and most of the trunk by 55 hours (Figure 4.37). The entire embryo is covered by 58 hours. Folding begins anterior to head (head fold of amnion; Figure 4.36), extends down either side of the embryo (lateral folds of the amnion; Figure 4.37) and at a later stage is initiated in the tail region (tail fold). The head fold folds back over head and the lateral folds proceed over the back and meet above the embryo (Figure 4.37). Once the folds meet, they fuse and the amnion and chorion become separate from each other (Figure 4.38). The cavity separating the amnion and chorion from each other is the **extraembryonic coelom**. Note that it is lined on all sides by mesoderm, and thus is, by definition, a true coelom.

The allantois develops as a diverticulum from the floor of the hindgut. The tissue that forms the allantois is endoderm covered by splanchnic mesoderm (i.e., splanchnopleure). The allantois expands rapidly and fills the extraembryonic coelom in the posterior part of the embryo. Expansion begins in the posterior part of the embryo and continues until the allantois has spread over the entire inner surface of the chorion. The allantois and chorion then fuse to form a common tissue layer, the **chorioallantoic membrane** (often called the **CAM**).

The yolk sac is the layer of extraembryonic splanchnopleure covering the yolk. The yolk sac forms by expansion of the blastoderm over the yolk. It is separated from other extraembryonic membranes (chorion, allantois) by the extraembryonic coelom.

b. Function of the extraembryonic membranes
1) **Amnion**: The fluid-filled amniotic cavity prevents dessication of the embryo and protects it from mechanical shock.

2) **Allantois**: The allantois serves as a reservoir for urine released into the cloaca by the mesonephric ducts. In addition, the allantoic splanchnic mesoderm becomes richly vascularized and the chorioallantoic membrane then serves as the respiratory organ of the embryo until hatching. Beginning a few days before hatching, the chorioallantoic membrane removes calcium from the egg shell. This calcium is deposited in the developing bones.

3) **Yolk sac**: The yolk sac is the digestive organ of the embryo until hatching. Breakdown of yolk and absorption of nutrients both occur in the yolk sac endoderm. Blood vessels of the yolk sac splanchnic mesoderm carry nutrients to the embryo. In addition, the yolk sac is the chief respiratory organ of the embryo prior to development of the chorioallantoic membrane.

64

C. Prepared slides of 72-hour embryo
1. WHOLE MOUNT
The visibility of the various parts of the 3-day embryo varies considerably in different specimens, with the best being found in more lightly stained embryos. It may be necessary to borrow slides from your neighbor to find all the structures mentioned. Be sure you can identify in the 72-hour embryo all structures described in the 48-hour whole mount, as well as the following new features.

a. Body shape
By 72 hours there is a second flexure of the body axis (Figure 4.39), this time at the **cervical** (neck) level near the posterior end of the **myelencephalon**. The anterior part of the body is curled almost completely around the heart. The head fold has moved further caudad and a new body fold, the **tail fold**, has now appeared at the posterior end of the embryo. The **tail** has lifted off the blastoderm by virtue of the development of the tail fold and is curled down and under as a result of a flexure in the tail.

b. Extraembryonic membranes
The amniotic folds have fused to form the **amnion** and **chorion**, both transparent membranes surrounding the body. The outline of these membranes can usually be seen dorsal to the myelencephalon. The organization of the amnion and chorion can better be seen in cross sections and will be examined later.

In addition to the amnion and chorion, a new extraembryonic membrane, the **allantois**, has formed just anterior to the tail. The allantois develops as an outpocketing of the hindgut. Depending on the exact age of your embryo, this membrane may be a very prominent, balloon-like structure or may be small and quite difficult to see. Ultimately the allantois grows large enough to line the entire shell, serving as a respiratory organ, in addition to its function as a reservoir for excretory products.

c. Endodermal derivatives
The **pharyngeal grooves** are easier to see at 72 hours than at 48 hours, and are most easily observed from the underside of the embryo. The **first visceral arch** is prominent as a bulge of

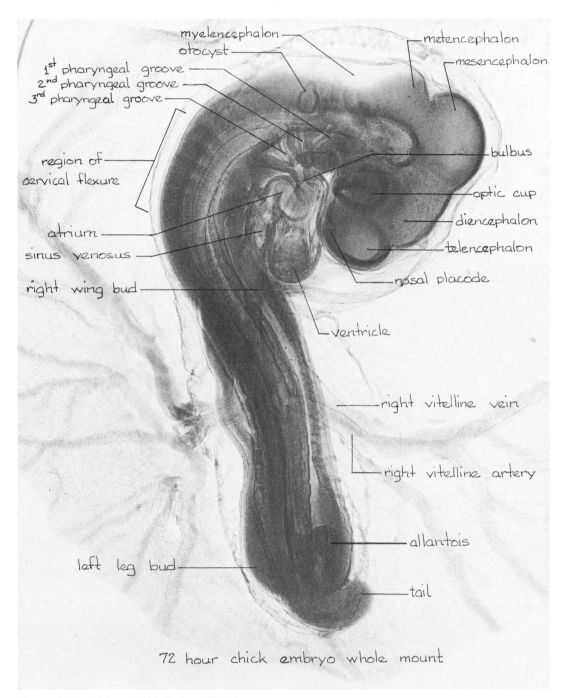

Figure 4.39. *72-hour Chick Embryo.* Whole mount.

tissue lying anterior and ventral to the slanted **first pharyngeal groove**. The first arch is split into two parts by a ventral inpocketing of head ectoderm, the **stomodeum**. The anterior part of the first arch (the **maxillary process**) will develop into the upper jaw, the posterior or **mandibular process** will develop into the lower jaw, and the stomodeum will develop into the mouth and mouth cavity. The anterior end of the foregut (**pharynx**) and the posterior end of

the stomodeum are in close contact at this stage, as will be seen in cross sections, but as yet the stomodeal and pharyngeal cavities are not in communication with each other.

Turn the embryo over and locate the **anterior intestinal portal** near the point where the large extraembryonic **vitelline veins** enter the body. In favorable specimens the floor of the foregut just ahead of the portal can be distinguished. In the tail region, locate the **posterior intestinal portal**. This has formed as a result of the posterior body fold or **tail fold**. Formation of the tail fold is similar to that of the head fold. The tail region lifts off the blastoderm and in doing so becomes tubular and surrounded by ectoderm (remember the analogy mentioned earlier, of poking your finger through a flat sheet of rubber?). The endodermal sheet forms a tube, the **hindgut**. Anteriorly, the opening of the tube forms the posterior intestinal portal. The allantois, mentioned above, is a ventral pocket of the hindgut.

d. Nervous system

The basic organization of the brain and spinal cord is similar to that in the 48-hour embryo. Locate the **telencephalon, diencephalon, mesencephalon, metencephalon** and **myelencephalon**. The telencephalon is now divided into right and left lobes which, together, will form the cerebrum of the adult. This division can be seen best if the slide is tilted at an acute angle so the anterior ends of telencephalon point upwards.

Development of sensory organs has advanced since 48 hours. The primordia for the sensory epithelial lining of the nose, the **nasal placodes**, are visible as pit-shaped thickenings of the ectoderm situated at the very anterior tip of the head. At a later stage some cells in each placode will differentiate into neurons (nerve cells) and project fibers to the telencephalon. Collectively, these neurons will make up the paired olfactory nerves (first cranial nerves). The **optic cup and lens** of the eye are also clearly differentiated.

e. Circulatory system

In good specimens, the aortic arches can be seen rather well. If they are not visible on your whole mount, borrow a neighbor's. The basic organization of the heart is the same as at 48 hours. Identify the **bulbus arteriosus, ventricle, atrium** and **sinus venosus**. The sinus venosus and atrium are seen best from the ventral side of the embryo. In good preparations, the **common cardinal veins** can be seen entering the sinus venosus when the underside of the embryo is examined. The vitelline veins enter the embryo

at about the same level that the vitelline arteries exist (best observed from the underside).

f. Limb buds

The **limb buds** are apparent as thickenings of the lateral body wall just anterior and six somite lengths posterior to the vitelline artery.

Now draw your 72-hour whole mount.

2. 72-HOUR CROSS SECTIONS

a. Quick survey

Once again, you should begin by scanning your slides with the dissecting microscope. It will also be helpful if you work out the angle at which your embryo was sectioned by determining the position of major landmark structures.

The first prominent feature usually encountered as you move posteriorly is the **myelencephalon**. Remember that this region of the brain will now be seen in frontal section because of the second body flexure. The myelencephalon is distinguished by its neuromeres, seen in previous embryos, and by the presence of the **otic vesicles** lying alongside. Proceeding along the row of sections, the **metencephalon** appears as a rounded distention connected to the widest part of the myelencephalon (Figures 4.40; 4.52/1). The **mesencephalon** lies anterior to the **metencephalon**, separated from it by the narrowed isthmus (see Figures 4.40 and 4.41). In subsequent sections, the walls of the mesencephalon are markedly circular in outline. Continue to follow the brain anteriorly by moving further along your serial sections. Remember the **diencephalon** is associated with the eyes (Figure 4.46) while the **telencephalon** is in the region where the **nasal placodes** invaginate as thickening of the epidermis (4.53/1)

Return now to the region where the mesencephalon assumes a circular form (e.g., Figure 4.42). The gaping space between the mesencephalon and the spinal cord is the cavity of the **pharynx**. Due to the bending or flexure of the embryo, the pharynx at this stage is also cut in nearly frontal sections (i.e., as though you are looking down into the pharynx from its roof). Find a section similar to the one shown in Figure 4.43 and identify the various visceral arches, pouches, and clefts. The **thyroid primordium** lies in the floor of the pharynx and will appear as a small circular epithelium-lined cavity lying between the second visceral arches (Figure 4.43). Find **Rathke's pouch** (Figure 4.44) opening off the mouth cavity and notice how the floor of the diencephalon (or **infundibulum**) pushes down to meet Rathke's pouch. The pituitary gland will form at the point where these two structures meet: the neural (posterior) lobe will form from the infundibulum while the anterior lobe will develop from the epithelial lining of Rathke's pouch.

Trace the pharynx posteriorly and notice how it changes into a narrow vertical slit. The dorsal part of the slit will become the **esophagus**. The ventral part is called the **laryngotracheal groove** and will become the trachea (Figures 4.45, 4.52/4). Trace the 1-t groove backwards until it bifurcates to form the **lung buds** (Figures 4.46, 4.50/1). The gut in this post-pharyngeal region hangs down in the embryonic coelom. Quickly move through the flattened midgut region tracing it posteriorly to where it again becomes tubular and forms the **hindgut**. Even further back, observe the **allantois** (Figures 4.49, 4.53/4), which appear as a ventral thick-walled diverticulum connecting to the hindgut.

b. Examination with the compound microscope

Examine the 72-hour cross sections a second time, this time with the compound microscope. Locate a section through the **pharynx** similar to Figure 4.42. Trace the **aortic arches** dorsally by passing forward in the sections. The aortic arches pass above the **pharyngeal pouches** and fuse with the paired **dorsal aortae** (i.e., Figure 4.41). Posterior to the **third aortic arch** the paired dorsal aortae fuse to form the unpaired dorsal aorta lying beneath the **notochord**. Blood is pumped from the heart up through the aortic arches, then carried to more posterior portions of the body via the dorsal aorta. Blood is returned to the sinus venosus from the (left and right) **vitelline veins** draining the yolk sac and the (left and right) **common cardinal veins**. Recall that the latter receive blood from anterior and poste-

rior regions of the body of the embryo via the **anterior** and **posterior cardinal veins**, respectively. The anterior cardinal veins can be located by their proximity to the **otocysts** (see Figure 4.40). Trace the anterior cardinals posteriorly until they fuse with the common cardinal veins (see Figure 4.46). Continue to trace the common cardinals into the sinus venosus. Locate the posterior cardinals and trace them back a few sections. Trace the sinus venosus posteriorly. A few sections behind the point where the common cardinals enter it (Figure 4.53/1), the sinus venosus becomes surrounded by darkly-stained spongy-looking tissue. This is the **liver primordium** (which will be examined in greater detail later). At the level where this liver tissue becomes apparent, the sinus venosus links with the **meatus venosus**,[13] a vessel that developed by fusion of the vitelline veins. Follow the meatus venosus posteriorly until it finally splits into the **vitelline veins**. Trace these veins out onto the yolk sac. The structure of the heart will be examined in detail in a later chapter.

Turn now to the eyes. Examine the **lens** under high power (400X). Notice that the lens is lined by a single layer of cells. Along the medial aspect of the lens the cells are very elongated and will later form fibers. Several of the cells in the section are undergoing mitosis: metaphase and anaphase figures are the most common stages seen. Later theses cells will stop dividing, elongate, and begin to synthesize crystalline proteins. These crystalline proteins (lens crystallins) will fill the lens cell cytoplasm and render the lens transparent. The optic cup contains tissue that will form both the **sensory (neural) retina** and the **pigmented retina** of the eye. The optic cup is connected to the diencephalon by a slender stalk of tissue, the **optic stalk**. The neural retina contains the cell bodies of the neurons responsible for light reception. As development proceeds, these neurons send axons that will project to the brain from the neural retina be following a groove on the outer lateral surface of the optic stalk.

---

13. The sinus venosus and meatus venosus can be distinguished as follows: The sinus venosus is the most posterior chamber of the heart, and has contractile properties. The meatus venosus is simply a blood vessel which, in the young embryo, allows blood returning from the yolk sac to bypass the liver capillaries and enter the sinus venosus. Later, as the meatus becomes invaded by liver tissue, it is transformed into the hepatic portal system. **67**

Proceed posterior in the embryo to examine the kidneys and somites. The prominent kidney of the 72-hour embryo, the **mesonephros**, is a system of tubules emptying into the **mesonephric duct** (formerly called the pronephric duct when the only kidney present was the pronephros; see Figures 4.49 and 4.50). By 72 hours of development, the pronephros has largely degenerated. The blood vessel lying directly above the mesonephric duct is the **postcardinal vein**. The detailed structure and function of the mesonephros will be treated in a later chapter.

The **somites** are condensed masses of mesodermal tissue on either side of the neural tube and notochord. In earlier stages, each somite was an epithelium surrounding a small cavity. By 72 hours the cavity has disappeared except at the dorsal margins of some of the more posterior (youngest) somites. At this stage the epithelium of the somite has segregated into three distinct regions: the **dermatome** (the epithelial portion of the somite lying just beneath the epidermis), the **myotome** (the epithelial layer spread along the ventral surface of the dermatome) and the **sclerotome** (the mesenchymal mass of tissue lying ventral to the epithelial components of the somite and adjacent to the spinal cord and notochord). The sclerotome was part of the original epithelial wall of the somite, but has undergone a transformation to become mesenchyme. At a stage somewhat later than 72 hours, the dermatome cells will also lose their epithelial arrangement, become mesenchymal, and migrate beneath the epidermis. Ultimately, the dermatome forms the deeper-lying tissue of the skin (dermis); the sclerotome forms the cartilaginous tissues of the spinal column; and the myotome forms the voluntary musculature of the trunk.

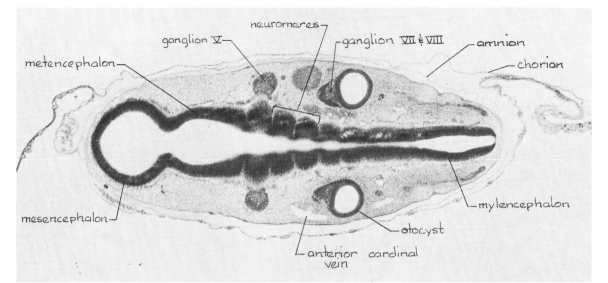

Figure 4.40

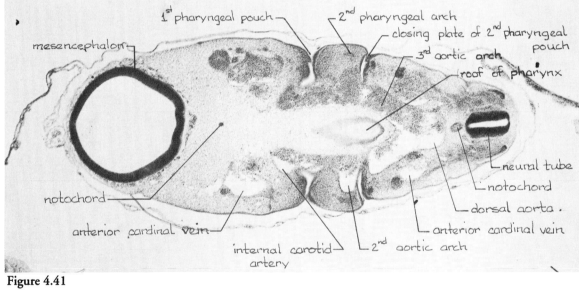

Figure 4.41

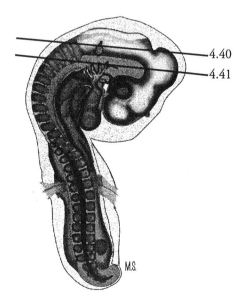

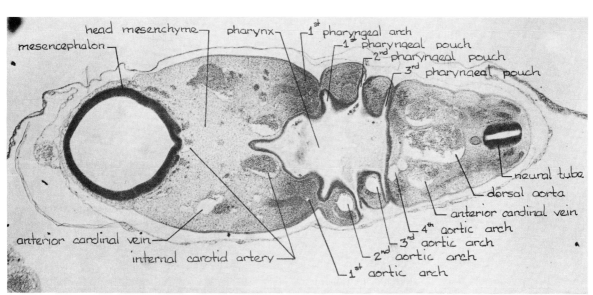

Figure 4.42

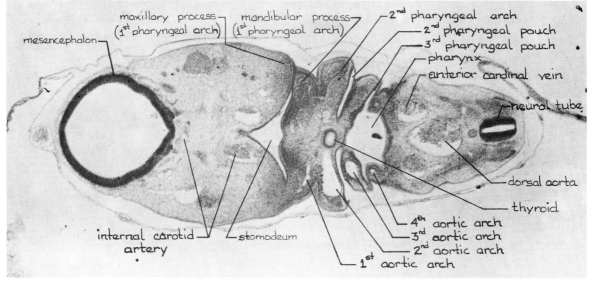

Figure 4.43

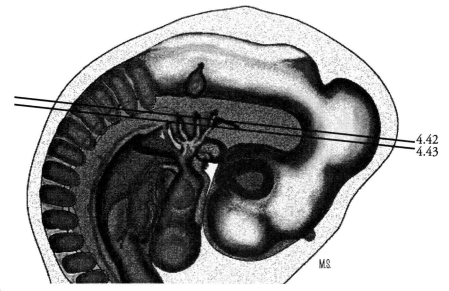

4.42
4.43

M.S.

70

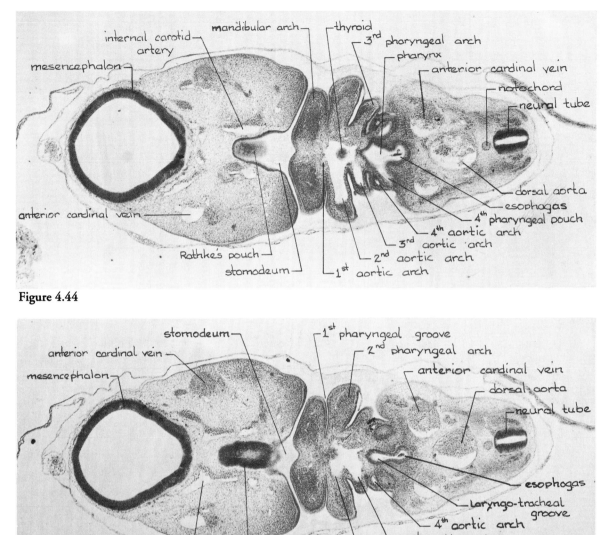

Figure 4.44

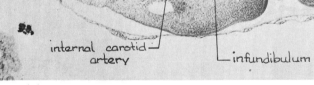

Figure 4.45

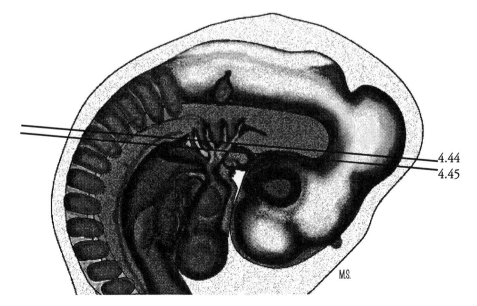

4.44
4.45

71

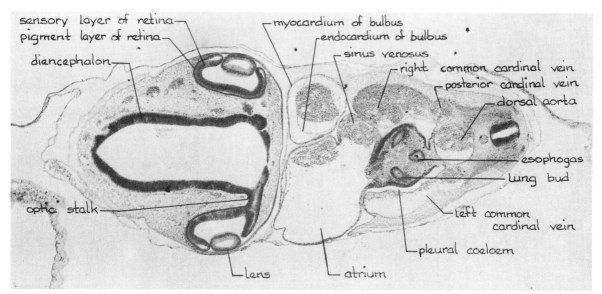

Figure 4.46

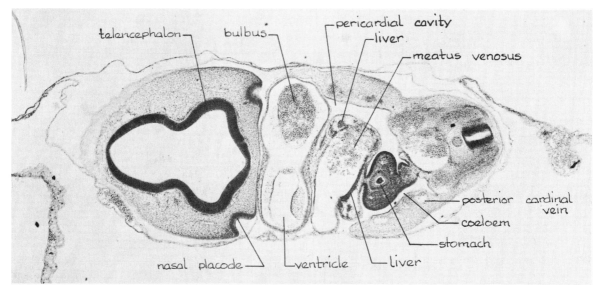

Figure 4.47

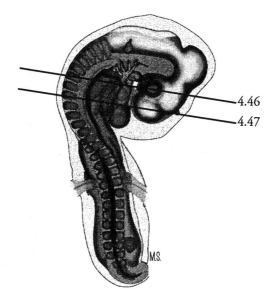

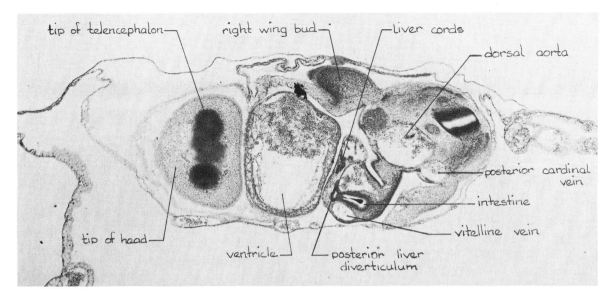

Figure 4.48

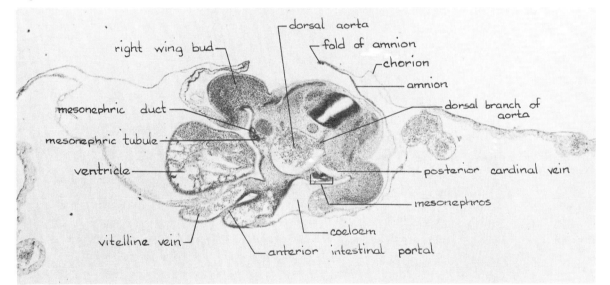

Figure 4.49

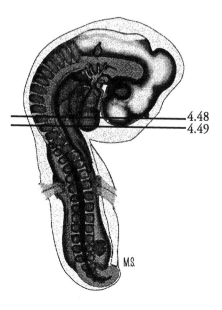

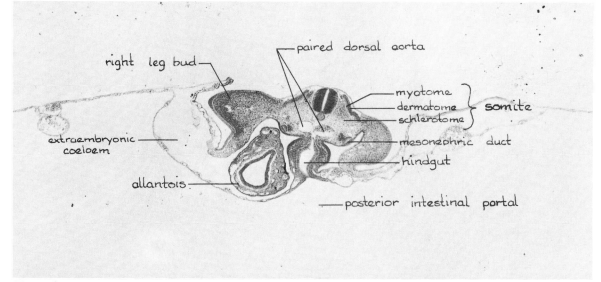

right leg bud

paired dorsal aorta

myotome
dermatome } somite
schlerotome

mesonephric duct

hindgut

extraembryonic coeloem

allantois

posterior intestinal portal

Figure 4.50

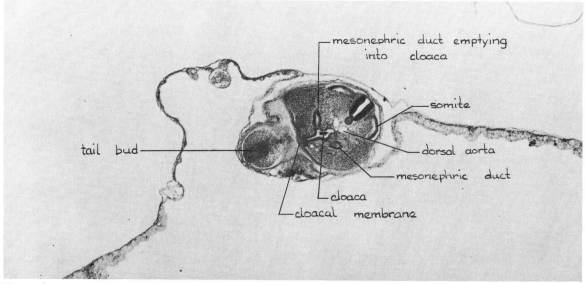

mesonephric duct emptying into cloaca

tail bud

somite

dorsal aorta

mesonephric duct

cloaca

cloacal membrane

Figure 4.51

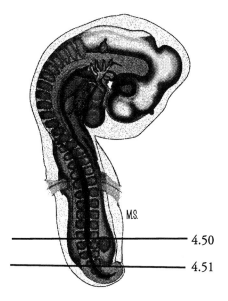

M.S.

4.50

4.51

Supplement to Chapter 4

The chick embryos in the micrographs that follow were prepared by fixing, paraffin embedding and sectioning in the same manner used for preparation of the slides used in the laboratory. However, sectioning was stopped part way into the embryo, then the remainder of the embryo was deparaffinated, and prepared for scanning electron microscopy. In all cases except Figure 4.52/2 and 4.53/3 sectioning was begun at the head and proceeded toward the tail.

Figure #: Structures illustrated
4.52/1: Metencephalon, neuromere, myelencephalon, otocyst, anterior cardinal vein, amnion, chorion, and yolk sac.

4.52/2: Surface view showing heart (ventricle), nasal placode, lens placode, pharyngeal grooves. Cut surface shows the anterior intestinal portal, vitelline veins, dorsal aorta, and posterior cardinals.

4.52/3: Diencephalon, optic cup, lens vesicle, aortic arch 1 (very small), aortic arch 2, thyroid, pharyngeal arches 2 and 3, paired dorsal aortae, anterior cardinal veins, notochord, spinal cord, and extraembryonic membranes.

4.52/4: Contrary to the usual case, this embryo is lying on his right side (rather than his left). Diencephalon, eyes, atrium (seen beneath the left eye in surface view), truncus, laryngotracheal groove, anterior cardinals, dorsal aorta.

4.53/1: Telencephalon, nasal placodes, bulbus, sinus venosus, cavity of atrium, common cardinals (right common cardinal emptying into the sinus venosus just beneath the plane of section), posterior cardinal, dorsal aorta, esophagus, lung buds.

4.53/2: Mid-sagittal section. Telencephalon, diencephalon with cavity of the right eye stalk, myelencephalon with neuromere, Rathke's pouch, stomodaeum, remnants of the oral plate, pharynx, pharyngeal pouches 1-4, bases of aortic arches 3 and 4, truncus, atrium, sinus venosus (opening of the right common cardinal is just visible in wall of sinus), meatus venosus showing strands of invading liver tissue, paired dorsal aortae above pharynx fusing into median dorsal aorta, openings of dorsal branches of aorta.

4.53/3: Sectioning was started at the tail and proceeded toward the head. Spinal cord, notochord, somites, dorsal aorta, posterior cardinals, mesonephric ducts, anterior intestinal portal, vitelline veins, wing bud, coelom, extraembryonic membranes.

4.53/4: Embryonic coelom forms triangular pouches on either side of cloaca. Lower part of hindgut is the allantois. Leg buds mesonephric ducts (visible above coelom cavities), dorsal aortae, and allantoic vessels.

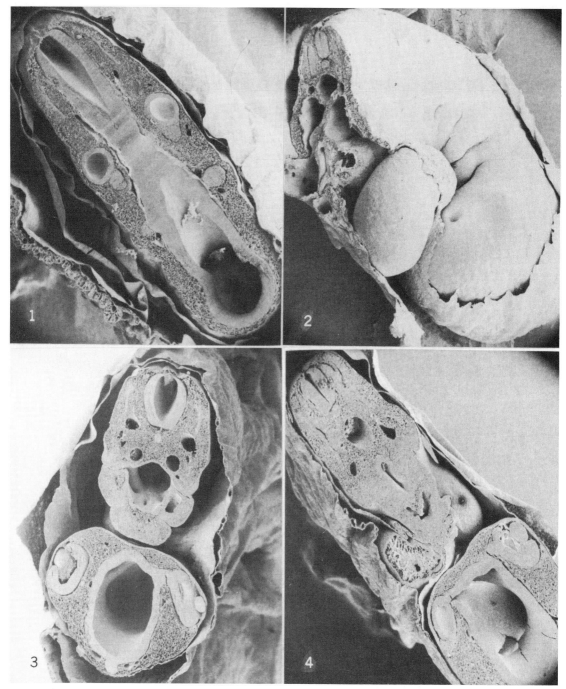

Figure 4.52

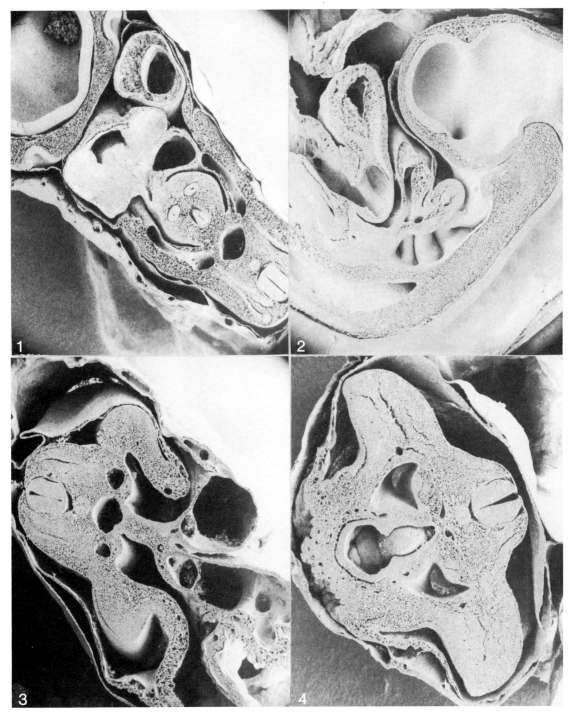

Figure 4.53

Observation of Living Embryos

I. INTRODUCTION

This chapter is devoted to laboratory observation and manipulations of living embryos. The objectives of these exercises are to provide an opportunity to acquire laboratory skills, and to reinforce the notion that development is a dynamic process, and not a series of static images on glass slides.

In the laboratory today you will be provided with living embryos that are approximately 33 hours, 60 hours, 5 days, and 10 days old. Your familiarity with the anatomy of the chick embryo will enable you to easily observe some of the noteworthy features of the living embryos, such as the heartbeat and the circulation of blood, movements of the embryos, and the extraembryonic membranes.

In addition to these general observations, two physiological features, both of significance to development, will be studied. The first important feature of development you will study today is the localization of actively developing embryonic regions, which exhibit a high rate of aerobic metabolism. This is because most, if not all, active developmental processes such as cell division and growth, morphogenetic movement, etc., are energy-dependent.

Localization of these regions of high metabolic rate is accomplished by immersion of the whole embryo in a buffered solution containing NADI reagent, which stains mitochondria. NADI contains two mild reducing agents (i.e., electron donors), α-*na*pthol and *di*methyl-p-phenylenediamine (hence the acronym NADI), which are oxidized by one of the cytochromes in mitochondria. The oxidized α-napthol and dimethyl-p-phenylenediamine then combine to form indophenol blue, a colored insoluble product. The areas of the embryo that have the greatest numbers of functional mitochondria will thus exhibit the most intense staining after treatment with the NADI reagent. As you examine the embryos after NADI staining, try to recall the developmental changes that are occurring at the most intensely stained region.

A second important feature of development to be studied in this laboratory period is the location of apoptotic (dying) cells in the normal embryo. Cell death plays an important role in the morphogenesis of certain organs. This phenomenon is most evident in the 5-day chick embryo. The regions of apoptosis can be identified by their affinity for certain stains such as **nile blue sulfate** or **nigrosin**. These stains are taken up only by dead or dying cells and are excluded from healthy viable cells.

You will determine the location of the regions of cell death in the 5-day embryo by immersing the whole embryo in a solution of nile blue sul-

79

fate. This dye will be taken up by cells that are in the process of dying. After making this identification, you should attempt to relate the observed staining patterns to the development of the structures in which they are found.

II. OBSERVATION OF THE 60-HOUR EMBRYO

By 60 hours of incubation the embryo is large and visible enough to allow handling with little trouble. This stage should, therefore, be examined first.

A. Opening the eggs and removal of the embryo

1). The embryo may be removed by handling the egg just as though you were cracking it into a skillet to fry it "sunny side up." Crack the egg against the edge of the lab bench as demonstrated by the laboratory instructor and allow the contents to flow into the pie plate.

2). If the yolk does not land with the embryo on top, you can rotate the yolk using the handle of a pair of large scissors in the area of the chalazae (the twisted cords of albumen extending from opposite poles of the yolk).

3). To remove the embryo from the yolk, place a filter paper ring on the blastoderm so that it sticks to the membrane surrounding the embryo. Then cut the area opaca with small scissors around the outside of the paper ring. The ring will not stick if the blastoderm is wet.

4). The ring, with the embryo attached, can be lifted with forceps to a dish of warm saline.

B. General observations

One of the striking features of this stage is the activity of the circulatory system. In the yolk

> **NOTE**
> Hurry through this section rather quickly so that the NADI staining (Part "C" below) can be begun as soon as possible. You will be able to make further observations of the general features of the embryo after placing it in the NADI stain.

sac, the numerous blood islands can be seen connecting with a network of capillaries that merge with the vitelline vessels. Observe the movement of corpuscles through the capillaries under your microscope (use either a compound or dissecting microscope). Can you distinguish arteries from veins in the extraembryonic circulation? Pay particular attention to the flow of blood through the heart, ventral aorta and aortic arches. Which part(s) of the heart contract(s) first? Locate the posterior margin of the amnion. Carefully tear the amnion with fine forceps so that the entire length of the embryo is exposed.

C. Localization of cytochrome oxidase activity

After making sure that most adhering yolk material has been washed away from the embryo, the staining procedure for cytochrome oxidase activity should be begun. The NADI reagent used to detect the activity of this enzyme must be prepared just before it is to be applied to the embryo. The two stock solutions have been prepared in advance.

NA solution: 0.01 M alpha-napthol (in 0.9% NaCl), combined with an equal volume of 0.066 M phosphate buffer, pH 5.8.

DI solution: 0.01 M dimethyl-p-phenylene diamine dissolved in 0.9% NaCl.

To make the active reagent, mix 20 ml "NA" with 10 ml "DI".

Pour the freshly mixed solution into a petri dish, then quickly transfer the embryo to this dish. Move the embryo by grasping the paper ring with forceps.

Watch for the first signs of color formation. Note the time when the embryo was placed in the solution and record in your notebook the areas that have stained blue at 3-minute intervals (for a total of 15-25 minutes).

Which part of the embryo is the first to stain? What does this suggest about the level of energy metabolism in that area? What other techniques could be used to confirm the results you obtained by this technique?

D. Shell membranes

Scrape the inner surface of the empty egg shells. The tough membranes lining the shell are the two non-living **shell membranes**. At the blunt end of the egg you can see a membrane that is stretched taut and does not contact the shell. This is the **inner shell membrane**. Puncture this membrane and observe the **outer shell membrane** in contact with the shell. The **air space** lies between these two membranes.

III. THE 5-DAY EMBRYO

A. Obtaining the embryo

Remove the embryo from the yolk using the same procedure as in the previous section and place it in a dish of warm Ringer's salt solution.

B. General observations

By this stage most of the general characteristics of body form have been established, and the limbs are becoming prominent features. How does the thickness of the ventricular wall of the heart compare with that of the 60-hour stage? Can you identify the **dorsal aorta**, **anterior** and **posterior cardinal veins**? Is the heart rate slower or faster than at 60 hours?

C. Cell death during embryogenesis

As discussed in the Introduction, the whole embryo will be stained to determine the size and location of naturally occurring populations of dead and dying cells. To stain only the naturally occurring apoptotic cells, it is important that the healthy viable cells of the embryo be kept alive and protected from accidental injury. The stain is dissolved in Ringer's solution, a physiologically balanced salt solution that will keep the embryo alive for the duration of the experiment. Try not to damage the embryo as you transfer it to the stain.

Transfer the embryo to a dish of warm Ringer's to wash away most of the adhering yolk. If the amnion is still intact, carefully tear it open with your watchmaker's forceps to expose the embryo. Then transfer the embryo to a dish containing nile blue sulfate in Ringer's (1:10,000 solution) and incubate for 15-30 minutes. The embryo will become tinted blue all over. Transfer the embryo back to a dish of Ringer's solution for about 30 minutes. The blue stain will disappear from all but the degenerating cells.

What centers of degeneration can be identified? Does this degeneration appear to be related to the shaping or formation of any particular structures at this stage?

IV. THE 33-HOUR EMBRYO

The embryo at this stage is quite small and transparent and difficult to see. To the naked eye, the embryonic region appears as a small whitish disk on the yolk; the embryo has the appearance of a white streak.

A. Staining the embryo

The 33-hour embryo may be stained with neutral red to allow visualization of structures without removing the embryo from the yolk. This may be done with the egg in a petri dish or finger bowl; the egg may be dry or in Ringer's solution. This allows observation of the embryo without disrupting the extraembryonic circulation, so the embryo should live longer.

Neutral red is a vital stain; it does not harm cells and stains only living cells. It is also used as a pH indicator; it is red at acid pH and yellow at basic pH, and localizes to acidic vesicles within cells.

Neutral red also acts as a selective dye. At low concentrations (about 10^{-4} M) only certain cells will take up the dye. Among the cell types selectively stained are neurons that contain biogenic amines (including serotonin, dopamine, norepinephrine), macrophages and neural crest cells. At higher concentrations, it will stain <u>all</u> living cells, but at different rates. It is not yet known why certain cell types selectively become stained, though they may contain an abundance of acidic vesicles.

In young avian embryos, neutral red will selectively stain neural crest cells under ideal conditions. The technique described below, however, will result in such selective staining only rarely. You should see most of the embryo stained, with darker staining in migrating neural crest cells. These will appear as dark dots, especially obvious in the rhombencephalic region, and along the dorsal neural tube.

1). Open the egg into a dish containing saline. The embryo will rotate to the top surface and you will be able to see the blastoderm, which will appear as a white disc about 1/4 to 1/2 inch in diameter. The embryo is in the center, but cannot be easily seen.

2). Stain the embryo by dropping a little (a drop or two) neutral red on it with a pasteur pipette. The stain will diffuse away if there is too much saline in the dish. Wait 3-5 minutes.

3). Since the vitelline layer is still between you and the embryo, it must be removed. The embryo should now be visible, after some of the stain is rinsed away, but will become much more obvious once the vitelline layer is removed. Drain away enough saline so that the embryo is not covered, then position on the blastoderm a paper ring cut just large enough to surround the embryo. Cut around the edges of the ring and remove the embryo to a petri dish filled with saline. Shake the embryo gently to loosen it from the ring. The vitelline layer should float away.

4). Note the regions of the embryo staining with the neutral red. Notice the brain regions, the number of somites (which identifies the stage of the embryo) and the area around the neural tube, which will contain the migrating neural crest cells appearing as dark dots.

B. Determination of cytochrome oxidase activity

Stain the embryo for cytochrome oxidase activity, using *freshly prepared* NADI reagent. Again, make observations at 3-minute intervals. What is the first region to stain? Are the highest activities found in the same regions as in the older embryos you have examined?

V. THE 9-10-DAY EMBRYO

A. Removal of the embryo

The extraembryonic membranes can be clearly identified at this stage. Carefully pick away as much of the shell as possible from the air-space end of the egg. Place a few drops of saline on the shell membrane, then very carefully expose the underlying **chorioallantoic membrane**.

Cautiously release the contents of the egg into a dish of warm saline solution.

B. General observations

Is any albumen present at this stage? How does the amount of yolk compare with earlier stages?

Cut through the chorioallantoic membrane. What cavity is now exposed? You can now see that the embryo is contained within another sac, the amnion.

Locate the umbilical cord. What large blood vessels run through the cord? The whitish loops within the cord are part of the intestine.

Cut out a piece of the yolk sac. What is the texture of the yolk itself? The sac has many villi or folds which increase its surface area. Why would increased surface area be of importance in the functioning of the yolk sac? How does the yolk reach the embryonic tissue?

C. Dissection of the embryo

Now we will dissect the embryo proper to complete a general survey of the embryo. Note that at this stage the embryo has digits and a tongue, feathers and even sunglasses!

Cut open the skull. How many lobes are there to the brain? Do you remember how the brain of the chick looked, even up to 72 hours of development? It was a hollow tube. But now, after the chick has been developing for 9 days, the brain is no longer hollow.

Next cut the lens out of the eye. As you do so take note of the structure of the optic cup and the layer of pigment lining it. Once you have removed the lens, place it on a piece of newsprint. Does the lens work?

Now cut open the body cavity. It is best to begin with an incision along the ventral midline. Find the heart, liver, kidneys, gallbladder, intestine, and bladder.

Finally dissect one of the limbs. Notice that all of the bones are already laid out. Can you find any muscles? As you have just seen, a 9-10 day old chick already has all its major body parts in place. During the next 11-12 days the main developmental requirement is growth.

VI. EXPLANT CULTURE OF CHICK EMBRYO

Culture of chick embryos in an "artificial shell" allows observation of the same embryo over time. Thus, the development of organ systems and extraembryonic membranes can be more easily studied. This "fourth dimension" of development is a concept difficult to teach with fixed materials or by observing embryos of several different ages. Many students have trouble visualizing the changes inherent in the development of the embryo. Like the ability to perceive three-dimensional form when looking at serial sections, visualizing development over time requires the integration of several different pictures. Unlike serial sections, looking at embryos at different stages does not provide all the information necessary, since there are gaps in the sequence of development. Therefore, what is visualized as the chick grows and develops may allow an accurate assessment of the actual changes that occur at intermediate times. Since embryos develop at different rates observing the same embryo is the best way to study development.

A. An introduction to teratology

Explant culture also provides an opportunity to study the effects of different substances on the development of the embryo. **Teratology** is the study of abnormal development. Research in this area includes the description of the morphology of specific defects, as well as an investi-gation of the mechanism that results in defects and the identification of causative agents, whether genetic or environmental. Although the literature regarding the descriptive aspects of teratology is extensive, there is little information on the mechanism of action of most **teratogenic agents**. Furthermore, in many instances, there is no information pointing to the causative agent is for a specific defect.

Much effort is presently being devoted to the identification of potential **teratogens** and to the elucidation of the mechanism by which birth defects are caused. One approach often used is to determine whether a substance is a potential teratogen and what its effects are on reproduction and pregnancy in laboratory animals. This approach is used by industry to establish guidelines to be followed in the commercial use of a specific product.

Another approach to research in teratology is to treat experimental animals with substances known to be teratogenic and to try to analyze the mechanisms by which a specific defect is caused. Examples of such substances are high (non-physiological) doses of adrenaline, sodium nitrite, and alcohol. Even when working with a known teratogen, with known dosages and known times of administration, it is difficult to determine the mechanism of action. There are many variables that can come into play, such as variations in the physiological condition of the maternal organism, that affect the outcome of the experimentation. Although the developing organism is susceptible to disturbances throughout embryonic and fetal life, individual organs are most sensitive to developmental insults during periods of rapid differentiation and growth (the **critical period**).

The field of teratology is a fascinating and often-ignored area of developmental biology. In today's laboratory you will have an opportunity to analyze the effects on development of treatment with a teratogenic agent. Using the background information on the agent provided by your laboratory instructor, and your observations, think of possible mechanisms to account for what you observe. Keep in mind that even when working with a known teratogen, the

83

actual frequency of positive results may be relatively low.

"We ought not to set them aside with idle thoughts or idle words about 'curiosities' or 'chances'. Not one of them is without meaning; not one that might become the beginning of excellent knowledge, if only we could answer the question—why is it rare or being rare, why did it in this instance happen?"—James Paget, 1882.

B. Alcohol as a teratogen

Maternal consumption of alcohol is believed to affect the mental ability and physical appearance of at least one in 750 infants born in the United States. Ethanol consumption is associated not only with abnormal live births but also with an increased incidence of stillbirths and a tenfold increase in perinatal mortality.

It was not until 1973 that the patterns of malformations that accompany alcohol abuse during pregnancy were fully characterized and termed "fetal alcohol syndrome." Major features of FAS are intrauterine and postnatal growth retardation, microcephaly (small brain), central nervous system dysfunction (including mental retardation and hyperactivity) and craniofacial abnormalities. These abnormalities include a narrow forehead, flat midface, narrow palpebral fissures (eyelid openings), a short nose, a long upper lip with a narrow vermillion border and a small or absent philtrum (the depression under the nose).

Although clinical and epidemiological studies in humans clearly show that alcohol affects development, the underlying developmental alterations were not understood until recently. In order to determine experimentally the critical exposure periods and the cellular affects of alcohol, experimental model systems not using humans had to be developed. When small doses of alcohol are administered to mice on day seven or eight of pregnancy (equivalent to three weeks in the human) the embryos develop craniofacial malformations closely resembling those seen in the human fetal alcohol syndrome. Recent studies by Katy Sulik and her collegues at the University of North Carolina suggest that the premature death of cells comprising the brain

and the neural crest cells, as well as inhibition of migration of the neural crest cells, can account for all of the physical deformities of this syndrome.

Alcohol has similar effects on chick development, so we will use the chicken embryo in our laboratory exercise as a model for fetal alcohol syndrome.

C. Instructions for "chick in a hammock"

1). First, clean your lab bench top using paper towels and 70% ethyl alcohol (EtOH). This will provide a semi-sterile area for you to work. Be careful to keep the area as clean as possible.

2). Collect the tools you will need: one pair of forceps (clean with alcohol), one paper egg tray, and one pie tin.

3). Disinfect an egg by wiping it gently, first with a paper towel moistened with 70% EtOH, and then with a towel dampened with Betadine solution. Let the egg air-dry by standing its wide end down in the egg tray. This causes the yolk to rotate so the embryo faces the egg's narrow end, away from the air space in the wide end of the egg.

4). Crack the wide end of the egg very gently, being careful not to push the shell into the egg. The object is to get just a crack in the shell, which you then use to start an opening through the shell. After cracking the egg, use the tips of your forceps to remove very small bits of shell at the large end of the egg, thereby exposing the air-space. Make the edges of the hole as smooth as possible so that the egg is not torn as it slips out the hole later. Make the hole as large as the air space, being careful not to puncture the shell membrane. Work very carefully. Be meticulous and work over a pie tin in case of disaster!

5). Hold the egg low over a dixie cup that has been fitted with a "hammock" of plastic wrap (your TA will provide these to you sterilized). Holding the egg with the small end up, carefully pierce the shell membrane with one point of your forceps, but <u>WITHOUT PUNTURING THE EGG CONTENTS</u>. You want to barely penetrate the shell membrane. This will dis-

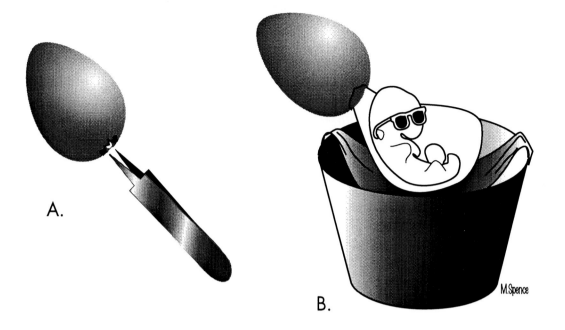

Figure 5.1. (A) To expel contents of egg, use fine forceps to remove small bits of the egg shell from the wide end of the egg to expose the shell membrane. Make a small puncture in the shell membrane. (B) Crack the narrow end to admit air which will release the egg contents through the window at the wide end of the shell. (Drawing after Cruz, 1993)

place the air space to the small end of the egg. A small amount of albumin may leak out, but the vacuum in the shell should keep the embryo in.

6). Make a hole in the small end of the egg by using a steel dissecting needle like a drill. As soon as you let in some air, the contents will flow out the hole at the bottom of the shell. If you make a hole, and the yolk does not come out, make the hole a little bit larger. If it still doesn't come out try gently rotating the egg on its long axis or piercing the exposed shell membrane in a different spot.

7). Add 0.5 ml Tyrode's solution with penicillin to the hammock using a sterile syringe and then replace the syringe in its case for reuse. Now add the alcohol solution; use 0.5 ml of 2% EtOH in Tyrode's, or 0.5 ml of 0.2% EtOH in Tyrodes', or 0.5 ml of the Tyrode's with no alcohol for control eggs. Since the volume of each egg is approximately 50 ml, this will give a final concentration of approximately 0.02% and 0.002% EtOH. Remember we are adding 0.5 ml of solution with penicillin, adding this to the

0.5 ml with EtOH gives [.5 ml 2% +.5 ml 0% = 1 ml 1%].

8). Place your embryo in the 37°C incubator on the shelf labeled "control", .02% or .002% EtOH.

9). Make daily observations of your embryo, and one from each of the other treatment groups until your next lab. The lab will be open according to the schedule provided by your TA. Take careful notes of your observations. They will be turned in with your write-up of this lab. Next week, all the data will be pooled for each lab, so that you can use a large data set for your analysis and your write-up.

You should pay particular attention to the development of yolk sac vascularization and blood circulation, comparing each concentration of EtOH to the control. Record your observations on a graded scale as illustrated below. Also, measure total embryo length and the length of a hind limb bud each time you observe your embryo. If your embryos live long enough, measure total embryo length and the length of a hind limb bud each time you observe your embryo. If your embryos live long enough, observe the development of facial structures, beak and palate. If your embryo dies, note the date and time you find it dead. Dispose of the egg in the garbage disposal, rinse the hammock, and throw it in the trash.

10). For your write-up think about the control we used for this experiment. Why was it necessary? For what were we controlling? The write-up should state a hypothesis, or pose a question, describe how you tested that hypothesis, and draw conclusions based on and supported by your observations and the class data. In your discussion you may speculate on possible effects the manipulation had on the embryos, and possible reasons the embryo failed to develop other than alcohol.

Example of a Grading Scheme:

++ Development strongly enhanced compared with control

+ Development slightly enhanced compared with control

0 No detectable effect (that is, indistinguishable from control)

– Development slightly impaired compared with control

– – Development severely impaired compared with control

– – – Development arrested or embryo killed within 24 hours

The Digestive System, Respiratory System, and the Body Cavities[1]

I. GENERAL CONSIDER-ATIONS

The embryonic rudiment of the digestive tract is essentially a tube of endoderm, the archenteron (primitive gut), which extends the length of the body. The archenteron has already been studied in an early phase of differentiation in the young frog embryo (Chapters 1 and 3). In amphibians the endodermal tube is produced directly by the invagination of the endoderm. On the other hand, in the amniotes it is formed after gastrulation by the appearance of head and tail folds, followed by the gradual apposition of the wings of the splanchnopleure to form a tubular endodermal passageway (see Chapter 4). So the embryo comes to possess a closed foregut, a midgut partly open into the yolk sac, and a closed hindgut. By definition, the **foregut** includes the liver diverticulum and everything anterior to it, the **midgut** represents the future small intestine, and the **hindgut** includes the point of origin of the large intestine and everything posterior.

Although most of the digestive tract is of endodermal origin, there is an ectodermal contribution where the openings to the exterior are to be established. At these points surface ectoderm becomes folded in, forming the ectoderm-lined **stomodeum** (mouth cavity) and **proctodeum** (anus). The major part of the mucous membrane of the adult mouth is thus of ectodermal origin, as well as components of the teeth and salivary glands.

The differentiation of the primitive gut varies in detail among different vertebrates, but the overall effect is the same: to produce a tract specialized to deal with the type of food eaten by the animal, and having a surface sufficiently large to hold, digest, and absorb as much food as the size of the body requires. Examples of specializations for dealing with a given diet include the development of heavy muscular walls to churn and move large masses of food, and the differentiation of enzyme-producing glands—both inside and outside the tract—to break down food particles chemically. Because of their relatively large body size, vertebrates have evolved a variety of devices to increase the absorptive surface in the digestive tube. Examples of such structures include the spiral fold in the intestines of elasmobranchs and primitive bony fishes, the blind pockets (pyloric caecae) in the stomachs of teleosts, the sac-like enlargement of stomachs generally, the elongated and coiled condition of the small intestine,

1. Material for this chapter was adapted from *Structure and Development of the Vertebrates*, 2nd ed., by Florence Moog. With permission of the author.

and the numerous tiny villi in the avian and mammalian intestines.

In addition to differentiating a functional digestive tract, the archenteron also gives rise to respiratory structures. All vertebrate embryos possess a visceral arch system in the pharyngeal region. In aquatic forms, these arches become functional gills: the mesenchyme (which is mesoderm and neural crest-derived) of the arch gives rise to the musculature, cartilage, and blood vessels of the gill, and the endoderm forms the thin membranous covering through which gas exchange takes place. In air-breathing vertebrates, the respiratory function is transferred from gills to lungs. Although the visceral arch complex of structures appears during embryonic life, they are modified into a variety of organs in the adult. Some of the mesenchyme-derived structures contribute to components of the hyoid apparatus; the endodermal structures contribute to the Eustachian tube (first pouch), the thymus and parathyroid glands (third and fourth pouches), and a part of the thyroid gland (fifth pouch). The bulk of the thyroid, also endodermal, comes from the floor of the pharynx.

The newly-developed lungs of higher vertebrates also derive from the pharyngeal endoderm. Most fishes possess some kind of an air sac that develops as a diverticulum from the ventral pharynx. In ancestors of the amniotes the air sac is believed to have served as an accessory respiratory organ, as it does in many living fishes, and thus became the forerunner of lungs. In amniotes, as the chick and pig embryos will show, the trachea and lung buds still originate from the ventral surface of the pharynx, and the digestive and respiratory systems always remain connected at the point where the tracheal diverticulum first appeared.

Since the digestive tract and its associated organs are derived from the archenteron, they are often referred to as being of endodermal origin. However the digestive and respiratory organs are not entirely endodermal. On the contrary, the bulk of most of these organs—the connective tissue, musculature, blood vessels—is mesodermal. In the case of the intestine, for example, only a very thin epithelial lining and the tiny

glands opening onto the lining surface are produced by the endoderm. All the rest is derived from the splanchnic mesoderm and surrounding mesenchyme.

The formation of the coelom has already been shown to result from a splitting of the hypomeric mesoderm into somatic and splanchnic layers with a cavity—the coelom—between them. With the closing of the gut, the splanchnopleuric layers of the two sides are brought together to form the mesenteries. The dorsal mesentery persists intact in most vertebrates, but the ventral mesentery largely disappears. The only structure that regularly retains a ventral mesentery is the liver, the development of which causes the mesentery ventral to the future stomach and duodenum to persist and differentiate into a lesser omentum (gut to liver) and falciform ligament (liver to body wall). The urinary bladder, where present, also retains a ventral mesentery.

The partitioning of the coelom is initiated in the embryo by the formation of the **transverse septum** which partially separates the anterior **pericardial cavity** from the posterior **peritoneal cavity**. In the 72-hour chick embryo, this separation begins where the common cardinal veins enter the heart. From this point a pleuropericardial membrane extends forward and back to isolate a ventral pericardial cavity. This stage of development is illustrated in the adult *Necturus* (Class Amphibia), which possesses a pericardial cavity and a pleuroperitoneal cavity (i.e., a common chamber dorsal and posterior to the heart, containing both lungs and abdominal viscera). Substantially the same relation persist in the reptiles and the birds.

In mammals two further developments occur. One of these involves a great expansion of the two pleural cavities. The lungs begin to develop in a thick mass of mesenchyme, known as the **mediastinum**, in which the esophagus and trachea are situated. As they grow, the lungs spread laterally into the body wall, splitting off a layer of parietal tissue before them. Eventually, the two pleural cavities meet in the mid-ventral line, though they remain separated from each other by a thin partition called the **mediastinal sep-**

tum. The heart is thus enclosed in a pericardial sac, which is composed of tissue split from the body wall of an earlier stage. The pericardial sac is entirely surrounded by the lung cavities.

The second development characteristic of mammals is the formation of a pair of pleuroperitoneal membranes, which wall off the posterior ends of the lung cavities. Thus the mammalian coelom becomes divided into four parts: two individual **pleural cavities**, a **pericardial cavity** enclosed by the pleural cavities, and a **peritoneal cavity**. The pleuroperitoneal membranes, together with some additional body wall tissue and muscle fibers derived from a pair of neck somites, are added onto the embryonic transverse septum to make up the mobile diaphragm, a structure peculiar to mammals.

II. THE DIGESTIVE TRACT AND ITS DERIVATIVES IN CHICK AND PIG EMBRYOS

NOTE ON PROCEDURE: Beginning with this laboratory period, the emphasis shifts from a general consideration of the form changes in the whole organism (morphogenesis) to a more detailed analysis of the morphological development of specific organs (organogenesis). To study the progressive changes in a developing organ, such as the liver or the pancreas, it is useful to examine those key stages in its development in which certain critical changes are apparent. Early primordia of many organs are seen, for example, in 33-hour or 48-hour chick embryos. More advanced stages of the same organ can be identified in sections of the 72-hour chick embryo and of the 10-mm pig embryo. By comparing the appearance of an organ in each of these stages, you can gain a good understanding of the structural changes involved in organ development. To aid you in your study of the more advanced developmental changes in individual organs, an atlas of representative cross-sections of the 10-mm pig embryo has been included (Chapter 10). Refer to it frequently in your studies of organogenesis.

In working with embryos from this point on, concentrate your attention on the system being studied. When you make your drawings, label only the structures under consideration.

A. 33-hour chick embryo, cross sections

Starting at the head end, identify the **oral plate** (Figure 4.14). Trace the pharynx back to the **anterior intestinal portal** (Figure 4.18). Does the pharynx have a dorsal mesentery? What represents the gut posterior to the anterior intestinal portal?

B. 48-hour chick embryo, cross sections

Find the oral plate (Figure 4.30). At each side the **mandibular arches** (future lower jaw) have now appeared. How has the oral plate changed in position since 33 hours? Examine the **mouth cavity**, or **stomodeum**, which is the confined space outside the oral plate; it is open at both sides. How many arches are present in your specimen? Identify the **visceral arch, pharyngeal pouches, pharyngeal grooves,** and **closing membranes**.

Locate the **anterior intestinal portal**. Is it in the same place as before? Is there a posterior **intestinal portal** present?

C. 72-hour chick embryo, cross sections

Examine the pharyngeal region and count the number of visceral arches. What criteria do you use to tell which arch is the first? Are any of the pouches open to the outside? In the floor of the pharynx, find a rounded body made of tall columnar cells; this is the beginning of the **thyroid gland** (Figure 4.43). Tracing back, note that the pharynx narrows down into the **laryngotracheal groove (l-t groove)** (Figure 4.45). The ventral portion of this groove gives off the **lung buds** (Figure 4.46). The mass of mesoderm containing the buds is surrounded by a restricted portion of the coelom, the **pleural cavity**. Below the pleural cavity is the heart, contained in the **pericardial cavity**. To what extent are these two subdivisions of the coelom separated from each other at this time? What structures appear to be playing a part in the separation of the pleural and pericardial cavities?

89

At the end of the lung region you will see that the sinus venosus, ventral to the gut, is bordered by small masses of darkly stained tissue. Proceeding posteriorly, you will find that these masses, which are the rudiments of the liver, are connected with the gut at the liver diverticulum (Figure 4.48). You may have seen the beginning of this diverticulum in the 48-hour embryo. In what direction does the diverticulum extend from its point of origin?

Find the posterior intestinal portal, and behind it the allantois bulging out of the ventral surface of the hindgut. The enlarged end of the gut is the cloaca, which in the bird becomes a common chamber for digestive, excretory, and genital systems (Figure 4.51). A little farther back identify the cloacal membrane, the point of union between the cloacal endoderm, and the outer ectoderm, where the opening to the outside is later established. Is there any gut posterior to this point?

D. 10-mm pig embryo
1. INTRODUCTION
You are familiar enough with the essentials of morphogenesis of the vertebrate body that we will not concentrate on the general morphogenesis of the pig embryo. Nevertheless, you will need to be able to orient yourself when studying various organs in the pig embryo. Here are some suggestions for attaining this orientation:

a. Model
A model consisting of representative cross sections mounted in proper sequence is on display in the laboratory. Examine this carefully; identify the major structures from your knowledge of frog and chick anatomy. You will find it helpful to refer to this model frequently during your studies of the pig embryo.

b. Survey
Survey your own serial cross sections quickly. Examine your set starting with the first slide. Be sure you can identify all the following structures: diencephalon; mesencephalon; myelencephalon; eyes; otocysts, pharyngeal (visceral) arches; lung buds and trachea; heart; liver; mesonephros; cloaca. You may find a dissecting microscope useful for this. Use the cross sec-

tions in Chapter 10 to aid you in identifying these structures.

2. ENDODERMAL STRUCTURES
Return now to the head region and find the first visceral pouch, which is the cavity of the future middle ear. The sections are often cut asymmetrically in this region, so concentrate on structures on one side of the embryo.

While it is not a part of the digestive system, it is convenient to point out here the components of the middle ear derived from the first pharyngeal pouch. Find this pouch in the 10-mm pig embryo. The dorsally directed extension of the pouch is the future tympanic cavity; the narrowed connection between the pouch and the pharynx is the future Eustachian tube (Figure 10.2). Trace the dorsal extension of the pouch. Does it extend all the way to the otocyst? In later development, the end of the pouch extension comes into contact with the otocyst, where it enlarges to form the middle ear cavity. The primordium of the future external auditory meatus (i.e., the canal from the tympanic membrane to the surface of the head) is the first pharyngeal groove. If the closing membrane of this groove is interrupted at this early stage, it will reform later. Continued growth carries the tympanic cavity well inside the head, so that contact between the pouch extension and the groove is temporarily lost. This situation is corrected, however, by growth of a diverticulum from the groove. This diverticulum establishes a new contact with the cavity, and at the point of this contact the tympanic membrane or eardrum is formed. Interestingly, the eardrum is constructed out of three germ layers: ectoderm from the groove, endoderm for the tympanic cavity, and mesenchyme that is caught in between.

Returning to the pharynx, identify the first and second pharyngeal arches (Figures 10.3 and 10.4). Locate the third pouch and note on its anterior face a dense mass of cells; this is an epithelial body or future parathyroid gland (Figure 10.4). Following the ventral wing of the third pouch backward, you will see that it extends for a considerable distance into the mesoderm below the pharyngeal floor, and eventually ends blindly. These ventral extensions of the third pouches are the future **thymus gland**. In passing, note the small dark **thyroid gland** in the mass of mesoderm; does it retain any connection with the cavity of the pharynx? Identify the **fourth pharyngeal pouches**, and the slender **laryngotracheal groove** stretching down from the mid-ventral line of the pharynx (Figure 10.6).

The ventral end of the groove splits off to form the **trachea**. Follow this until its distal end divides into the two **bronchi** (Figure 10.12), at the tips of which are the **lung buds**.[2] Examine the buds under high power. How do the endoderm and mesenchyme differ in appearance?

Is the **pleural** region of the coelom separated from the **pericardial** region? Examining the relation between the lung cavity and the abdominal cavity, determine whether it would be proper to speak of a **pleuroperitoneal** cavity at this stage, or whether the two regions are completely separated. Find the first appearance of the **transverse septum** as a broad band of tissue ventral to the lung buds (Figure 10.13). Trace it back to see that it forms a slanting wall behind the heart. Note that a considerable amount of **liver** tissue is embedded in the septum.

Identify the **esophagus** (Figures 10.7, 10.8), and find where it widens into the **stomach** (Figures 10.14, 10.15). Where does the dorsal mesentery first appear? Examine the stomach under high power. Can you differentiate between the future epithelial covering of the mucosa, which is the single layer of columnar endodermal cells, and the future connective tissue and musculature (mesenchyme)? Ventral to the stomach is the liver.

Posterior to the stomach find the entrance of the **common bile duct** into the intestine (Figure 10.16). This is the site of the original liver diverticulum. Within the liver substance, identify the small **gall bladder**, which sends a duct into the common bile duct (Figures 10.16, 10.17). In the mesenchyme surrounding the intestine, find the two masses of glandular tissue

which will develop into the **dorsal** and **ventral** pancreas (Figure 10.17). Find also the duct of each of these bodies. Does the ventral duct enter directly into the gut? In the pig, the dorsal duct becomes the functional duct, but in man the ventral duct becomes functional, although the dorsal connection usually persists as the accessory duct of Santorini.

Trace the gut posteriorly. What course does it take in the umbilical cord? At the base of the tail, note the **rectum** (Figure 10.19) entering the dorsal surface of the cloaca. The **cloacal membrane** is close by (Figure 10.19). A little anterior to the membrane the **allantoic stalk** will be seen to originate from the ventral surface of the cloaca. Actually, the cloaca is already becoming subdivided into the rectal portion and a urogenital portion, the allantois having its connection with the latter. Trace the allantois into the umbilical cord (Figure 10.12 through 10.15).

---

2. The major lobes of the left lung of the adult are formed as a result of a bifurcation of the lung bud at the tip of the left bronchus. A similar bifurcation occurs at the caudal end of the right bronchus to form two of the three major lobes of the right lung. In addition a third lobe (the upper lobe) of the right lung develops in the pig (and also in humans). In the 10-mm pig, this lobe appears as a rounded diverticulum that bulges off the right side of the trachea anterior to the point where the trachea divides to form the bronchi.

linked physically in the adults of fish and amphibians, but are largely separate in the mammals, although some degree of anatomical connection persists.

II. EMBRYOLOGY OF THE UROGENITAL SYSTEM

Development of the urogenital system begins with the segregation of the mesoderm into dorsal, intermediate, and ventral sections. The **intermediate mesoderm—mesomere or nephrotome**—gives rise to almost the whole of the urinary and genital apparatus. Only the cloaca and structures which are derived from it in mammals, such as the bladder, the urethra, and a part of the vagina, are of endodermal origin.

Differentiation of the mesomere starts at its anterior end and proceeds posteriorly. This differentiation involves the formation of tubules that collect nitrogenous wastes from the blood and carry them to a duct through which they are passed out of the body.

In most embryos, the anterior portion of the mesomere (fifth to fifteenth somites in the chick) differentiates into the **pronephros**. The mesomere becomes segmented, and each segmental block gives rise to a single pronephric tubule. These tubules open into the coelom via **nephrostomes**, and have an indirect connection (**external glomerulus**) with the arterial blood system. The **pronephric** duct arises as a communication passage from one tubule to the next. It continues to grow posteriorly from the last tubule and eventually breaks an opening into the cloaca.

The pronephros is the working kidney of the anamniote (fish and amphibian) embryo. Waste molecules circulating

THE UROGENITAL SYSTEM[1]

I. INTRODUCTION

The excretory and reproductive systems of vertebrates are generally discussed together because they are structurally and developmentally related. The excretory organs begin to form very early in the embryonic period, since a special waste-removing apparatus becomes imperative as soon as the body grows too large for wastes to diffuse directly away. On the other hand, the gonads, possibly because they are active only in the adult period, make their appearance relatively late in embryogenesis. They develop on the kidney surface and gradually appropriate some of the first-formed excretory apparatus, which are converted to their own use. The two systems are thus intimately associated during early development. However, the extent to which this association persists into the adult stages varies inversely with the position of the species on the vertebrate evolutionary scale. The two systems are closely

1. Edited from *Structure and Development of the Vertebrates*, 2nd ed. (Abridged), by Florence Moog. With permission of the author.

in the capillaries of the glomerulus are filtered into the general coelomic fluid, and some of this fluid is then drawn into the nephrostomes. From there it can be eliminated by the pronephric tubules and duct. The pronephros does not function in amniote embryos, however, and consequently the differentiation of the pronephric tubules in amniotes is quite variable. The tubules soon degenerate and disappear. The reason they develop at all may be that they are needed to initiate the development of the pronephric duct.

Posterior to the pronephros the mesomere remains unsegmented, and surrounds the pronephric duct. Eventually, this tissue produces a series of **mesonepheric tubules**, starting at the pronephros and gradually proceeding posteriorly. These make openings into the pronephric duct, which is then known as the **mesonephric or Wolffian duct**. The **mesonephric tubules** differ from those of the pronephros in many respects. First, they are connected to the bloodstream through internal glomeruli contained within the ends of the tubules themselves. Second, nephrostomes either do not form or are transitory, so that there is no effective opening into the coelom. Third, several tubules occur per body segment. Finally, the mesonephros is a much longer structure than the pronephros.

The mesonephros is the functional kidney of the adult anamniote and of the embryonic amniote. In the amniotes it is superseded, late in the developmental period, by a third kidney, the **metanephros**. The metanephros develops from the most posterior region of the mesomere. It begins as a diverticulum from the mesonephric duct. This diverticulum, the **ureteric bud**, grows forward and makes contact with the still undifferentiated end of the mesomere. This mesomere tissue, now called the metanephric blastema, wraps around the end of the ureteric bud and the two together differentiate into the mature kidney, the bud developing into the ureter and system of collecting ducts, and the blastema giving rise to the urine-producing tubules. The metanephric tubules resemble those of the mesonephros in having internal glomeruli.

The entire mesomere is outside the coelom, and is so placed that the developing kidneys become covered by a layer of parietal peritoneum. At a relatively late stage in the differentiation of the mesomere, a part of this covering peritoneum thickens to form the genital ridge, from which the gonads develop.

III. UROGENITAL STRUCTURES IN CHICK AND PIG EMBRYOS

A. 33-hour chick embryo

A pronephros has begun to appear in the chick embryo by 33 hours. It occurs in the fifth to fifteenth segments only. As is generally true in amniotes, this first kidney is nonfunctional, and it is less well-differentiated than in some mammalian embryos.

In the cross sections find the first somite at the posterior end of the heart, and count back eight or more somites. Here find a small ovoid mass of cells lying between the somite and the lateral plate (hypomere). This mass is the nephrotome (mesomere) (Figure 4.19). It is rarely organized into actual tubules in the chick, but a **pronephric duct** is usually present, lying on the dorsal surface of the nephrotome. The duct grows back beyond the pronephric region and eventually opens into the hindgut (at about 60 hours).

B. 48-hour chick embryo

At about the level of the anterior intestinal portal, find the posterior cardinal veins to each side of the aorta. Follow one of these veins back until you find a small, thick-walled duct ventromedial to it. This is the pronephric duct, which is now called the **mesonephric (Wolffian) duct** (Figure 4.35), since the mesonephros has begun to develop around it.

The beginnings of the **mesonephros** will be seen as small masses of nephrogenous tissue, derived from the nephrotome, clustering around the duct. Over a considerable length of the body, mesonephric tubules have already differentiated from part of the nephrogenous tissue. The duct is lateral to the tubules. Find a **mesonephric tubule** opening into the duct. Examine the first

94

two or three tubules and see whether you can find one with a **nephrostome**, or opening into the coelom. Nephrostomes are characteristic of pronephric tubules, which do not have internal glomeruli, but the first few mesonephric tubules also generally possess a coelomic aperture.

Follow the mesonephros back and observe that it is a very long structure. Toward its posterior end, tubules are not yet formed. How far does the duct reach?

C. 72-hour chick embryo

Identify the **mesonephric duct** and a **mesonephric tubule** (Figure 4.49). Note that a considerable amount of undifferentiated nephrogenous tissue is still present alongside the tubule. This tissue will also differentiate into tubules, for the mesonephros is characterized by the presence of three or more tubules within each segment of the body. The posterior cardinal vein is still present dorsolateral to the kidney, and a smaller **subcardinal vein** has now appeared on the ventral side. Note that the mesonephros bulges slightly into the coelom.

Follow the duct beyond the kidney and find where it opens into the cloaca (Figure 4.51). The mesonephric kidney begins to function during the fifth day of incubation in the chick.

D. 10-mm pig embryo

The mesonephros is fully differentiated and functional at the stage represented by the 10 mm pig. Its enormous size is correlated with the epitheliochorial type of placenta found in the pig. In placentas of this type, several layers of tissue are interposed between the fetal and maternal blood streams, and nitrogenous wastes from the embryonic body are not removed very efficiently. Instead, the embryo develops a massive mesonephros and carries on its own excretion.

Choose for study a section through the widest part of the mesonephros, where the two kidneys are separated from each other by a large blood sinus, the **subcardinal anastomosis** (i.e., union of the two subcardinal veins; Figures 10.19 through 10.21). The major part of the structure is made up of the coiled mesonephric tubules,

which are cut at various angles. The mesonephric duct is at the ventral edge of the kidney. Find a tubule opening into it.

Medially note the large **renal** (**Malpighian**) **corpuscles**. Each of these is made up of a thin-walled **Bowman's capsule** surrounding a knot of capillaries which comprise the **glomerulus** (Figure 10.16). The capsule, which leads into a tubule, is a two-layered structure with an intervening lumen, but its lumen generally disappears from shrinkage when sections are prepared. Find the connection (lateral or **renal artery**) between a glomerulus and the dorsal aorta.

The **posterior cardinal vein** is situated on the dorsal side of the mesonephros. It is beginning to disappear at this stage, and may not be present in every section. Capillary connections exist between the posterior cardinal and the **subcardinals**.

The **gonad** has begun to make its appearance in the form of the **genital ridge** (Figure 10.19), a thickening of the parietal peritoneum covering the medial side of each mesonephros. Examine the ridge and determine how far it extends anteriorly and posteriorly.

To study the relation of the urogenital system to the cloaca, find the tip of the tail. Proceed forward in the tail (i.e., posterior in the series of sections) until you come to the **cloacal membrane**, where the endoderm of the cloaca meets the ectoderm of the body surface (Figure 10.17). Proceed as follows: to the cavity of the cloaca; to the point where the rectum enters the cloaca (and note whether the entrance is on the dorsal or the ventral side); to the place where the cloaca bears lateral extensions which receive the mesonephic ducts; to the place where the mesonephric ducts enter the cloaca; to the point of entrance of the **allantoic stalk** (and note whether this is dorsal or ventral). Follow the allantois into the umbilical cord.

Returning to the cloaca, follow one of the mesonephric ducts as it runs forward alongside the rectum. Locate the point where the duct gives off a small sprout (on its dorsal or ventral side?) that continues forward independently. This sprout is the **ureteric bud**, the rudiment of the third or **metanephric kidney**. Follow it until you find its enlarged distal end surrounded by condensed mesenchyme. The stalk of this structure is the future **ureter**, the enlarged end is the future **renal pelvis**, and the enveloping mesenchyme, the **metanephric blastema**, is the future **cortex** (Figure 10.21). The blastema differentiates into metanephric tubules, whereas the epithelial wall of the pelvis gives rise to the system of collecting ducts (**medulla**) of the mature kidney.

Finally, return to the mesonephric duct and trace it into the mesonephros.

THE CIRCULATORY SYSTEM

I. THE EARLY DEVELOPMENT OF THE CIRCULATORY SYSTEM[1]

The circulatory system of the early amniote embryo closely resembles that of the adults of primitive vertebrates. Blood exits from the heart and ventral aorta into a series of paired aortic arches, and thence to the dorsal aorta. As development continues, this primitive configuration becomes progressively altered, with the loss of several of the aortic arches and extensive modification of the remainder. The circulatory system is one of the first organ systems to become functional in the amniote embryo.

Primitive circulation is initiated with development below the foregut of a **tubular heart** from splanchnic mesoderm. With the appearance of this primitive heart tube and system of peripheral blood vessels an early circulatory system loop is established, allowing blood to flow from the embryonic to the extraembryonic circulation. Anteriorly the heart continues as the **ventral aortae**, which extend anteriorly and curve dorsally around the front of the pharynx. These dorsally directed loops, which become incorporated into the mandibular arches, are the first **aortic arches**. Above the pharynx they join the dorsal aortae, which gradually grow back to the tail. Midway down the embryo, large branches of the dorsal aortae, the **vitelline arteries**, carry blood to the yolk sac. Blood returns from the yolk sac to the embryonic circulation via the paired **vitelline veins**. This is true even in mammals, in which the yolk sac is vestigial.

While these processes are going on, the **cardinal veins** are laid down in the body wall on each side of the dorsal aorta. These veins connect with the posterior end of the heart by way of the ventrally directed **common cardinal veins**, or **ducts of Cuvier**. Like the heart, all these vessels are formed by the assembly of individual angioblasts into vascular cords. This is known as the process of vasculogenesis. Only later do new vessels appear by sprouting from pre-existing vessels, which is the process of angiogenesis.

The most striking change occurring soon after the establishment of circulation is the appearance of the aortic arches. These are laid down in anteroposterior sequence until six are present (Figures 8.1, 8.2, 8.3). Each one is, like the first, a connection between the ventral and dorsal aortae. Very early, however, the first, second, and fifth arches disappear in embryos of almost all land-living species (Figure 8.1, 8.2, Table 8.1).

1. The material in part I. was modified from *Structure and Development of the Vertebrates*, 2nd edition, abridged, by Florence Moog. With permission of the author.

c h a p t e r 8

97

Table 8.1. Adult Derivatives of the Aortic Arches.

Arch #	Adult Derivative
1	disappears
2	disappears
3	common carotid arteries; part of the internal carotid arteries
4	right: right subclavian artery (part)
	left: ascending aorta (part)
5	disappears
6	ductus arteriosus and (after birth) ligamentum arteriosum
	pulmonary arteries

The **posterior cardinals** early in their development are closely associated with the kidneys, but they are soon superseded by a pair of **subcardinals,** which develop ventromedially to the kidneys. These new vessels extend forward and unite with the anterior ends of the posterior cardinals; thereafter a portion of each posterior cardinal degenerates, so that blood flowing up from the tail region in the posterior cardinals must detour through capillaries in the kidneys and then travel on to the heart by way of the subcardinals. This system of venous vessels and capillaries forms the **renal portal system.**[2]

The development of the **hepatic portal system** is initiated by the growth of the liver immediately behind the heart. Because of this position, the expansion of the liver has a profound effect on the development of the vitelline veins. In the chick embryo, the two veins become fused within the liver and for some distance beyond, into a single vessel called the **meatus venosus.** The meatus venosus is invaded by cords of developing liver tissue until it is reduced to a meshwork of blood sinusoids. Gradually the sinusoids are separated into two connected capillary beds: one capillary bed drains the distal parts of the vitelline veins, and this receives blood from the

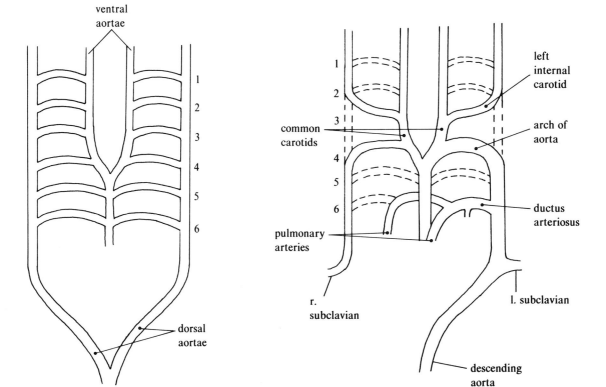

Figure 8.1. *Development of the six aortic arches.*

Figure 8.2. *Changes in the six aortic arches during development.* The dotted lines represent vessels that totally disappear. See text for additional details.

2. A renal portal system is found in vertebrates with a functional mesonephros in adult life. It has been presumed to function as a second filtering system for blood in these lower forms, to compensate for the relatively low vascular pressure and less efficient mesonephric kidney. However, physiological evidence to support this view is not forthcoming.

yolk sac; the other supplies the proximal part of the meatus, which transmits blood to the heart. This is essentially a **hepatic portal system**. Other veins, primarily from the digestive organs, also feed into this system. After hatching, the extraembryonic parts of the vitelline veins disappear, but the system continues to function because the mesenteric vein, which drains the gut, continues sending blood through the liver sinusoids.

In mammals the development of the hepatic portal system is fundamentally the same as in birds, except that the vitelline veins do not become fused. Instead, they are broken up directly into capillary beds by the growth of the liver (Figure 8.4).

An open channel, the **ductus venosus**, also passes through the liver in the mammalian embryo. This channel develops as a result of the invasion of the liver by the umbilical veins. The **right umbilical vein** becomes reduced early in the embryonic period, but the left carries a large stream of placental blood that forces a right of way directly through the liver substance. The ductus venosus serves the same function as does the avian meatus venosus, but the two differ in fate as well as in origin. Whereas the meatus is split into capillaries at an early stage, the ductus

remains open until birth, only collapsing when the placental circulation ceases.

II. DEVELOPMENT OF THE HEART

The heart begins beating very early in development. In the chick embryo, which hatches on the twenty-first day after incubation, the heart begins beating by thirty hours of development. In humans, with a nine-month gestational period, the heart begins to beat by the end of the third week. The onset of function does not interfere with the laying down of new vessels, which goes on continuously.

In your earlier survey of the chick embryo, you observed that development of the heart begins with the formation of two endocardial tubules from splanchnic mesoderm. These endocardial tubules fuse early on to form a primitive heart, connected to the rest of the circulatory system posteriorly by vitelline veins and anteriorly by the ventral aortae. The primitive heart tube quickly expands, and soon several primitive chambers are present: the **sinus venosus**, the **atrium**, the **ventricle**, and the **bulbus cordis**. The ventricle and bulbus cordis continue to

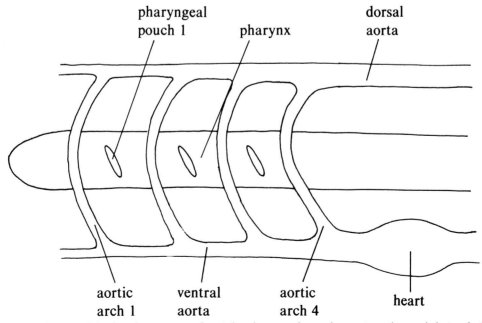

Figure 8.3. *Side view of the first four aortic arches.* This diagram shows the aortic arches and their relationship to the pharyngeal pouches. The aortic arches are embedded in the mesenchyme of the pharyngeal arches.

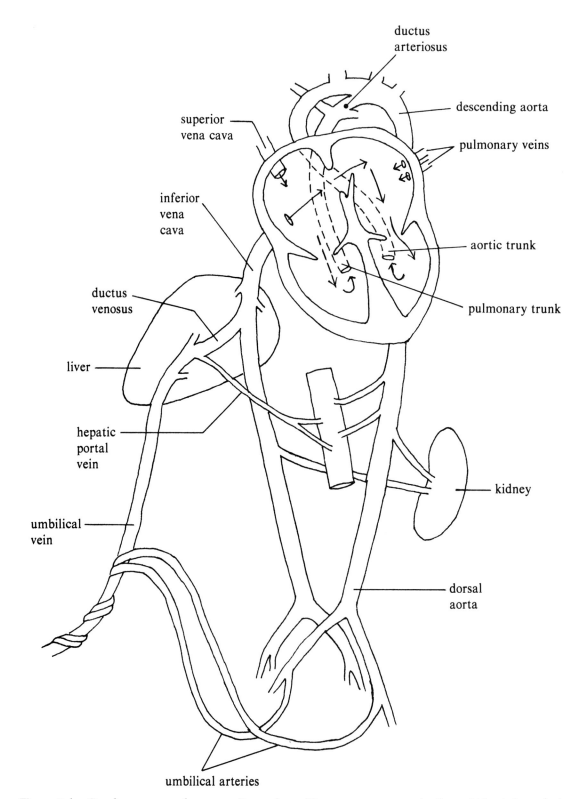

ductus arteriosus

descending aorta

superior vena cava

pulmonary veins

inferior vena cava

aortic trunk

ductus venosus

pulmonary trunk

liver

hepatic portal vein

kidney

umbilical vein

dorsal aorta

umbilical arteries

Figure 8.4. *Circulatory system of a mammalian embryo.* The arrows represent the flow of blood through the heart in the embryo.

develop at a greater rate than the remaining chambers, soon giving rise to the S-shaped heart seen in the 48-hour embryo (Figure 4.24, 4.26). One result of this differential growth is the shifting of the ventricle posteriorly in relation to the atrium, thus placing the two main regions of the heart in their proper adult position.

Two masses of tissue begin to grow in the region between the atrium and ventricle and begin narrowing the **atrioventricular canal** (Figure 8.5). In mammals, these two masses, called **endocardial cushions**, normally meet and fuse by the 10-mm stage. In the atrium, separation begins with the extension from the dorsal wall of the atrium of the **interatrial septum** (**septum primum**) which grows downward toward the endocardial cushions (Figure 8.5). Prior to reaching the cushion, an opening forms in the septum to form the **interatrial foramen**. Thus, the septum is never totally closed. A second septum then begins to form on the right side, adjacent to the first interatrial septum, and grows down toward the endocardial cushion, but only extends to just beyond the lower edge of the second interatrial foramen. The two partitions form an oval-shaped foramen, the **foramen ovale** (Figures 8.4, 8.5). The foramen ovale allows a large part of the blood entering the right atrium to flow directly to the left atrium and thus bypass pulmonary circulation. At birth, with the establishment of pulmonary circulation, the pressure of blood returning from the lungs to the left atrium forces the foramen shut. There is eventually adhesion of the two septa, so the foramen is effectively obliterated in adults.

The ventricle is partitioned into right and left sides by growth of the **interventricular septum**. At the same time that the interventricular septum develops, the bulbus cordis is partitioned into **pulmonary** and **aortic trunks**. This sets the stage for blood to flow from the right ventricle into the lungs via the pulmonary trunk, and from the left ventricle to the systemic circulation via the aorta. During embryonic and fetal life, blood flow to the pulmonary arc is largely bypassed by the foramen ovale, discussed above,

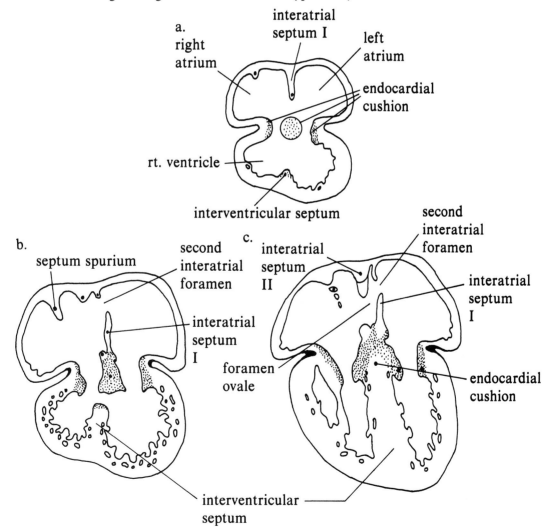

Figure 8.5. *Development of the heart in a mammalian embryo.* A. 7-mm embryo. B. 9-mm embryo. C. 30-mm embryo.

101

and by the **ductus arteriosus**, which connects the pulmonary trunk to the aortic trunk (Figure 8.4). This bypass is eliminated at birth, with closure of the ductus arteriosus (see below).

III. CARDIOVASCULAR CHANGES AT BIRTH

With separation from the maternal circulation and breathing, several changes take place in the newborn circulatory system. These are summarized in Table 8.2 and are briefly discussed below.

Table 8.2. Adult derivatives of embryonic and fetal circulatory structures.

Embryonic Structure	Adult Derivative
foramen ovale	closes
ductus arteriosus	ligamentum arteriosum
umbilical vein	
•ductus venosus segment	ligamentum venosum
•umbilical vein	ligamentum teres
umbilical arteries	
•proximal portion	superior vesical arteries
•distal portion	lateral umbilical ligaments

As discussed in the preceding section on the development of the heart, the foramen ovale closes under the increased pressure in the left atrium. It is estimated that 25% of the adult population have a foramen ovale in which, although closed, the septa have not adhered and is thus potentially patent (and thus a potential source of trouble).

The ductus arteriosus, mentioned above, also closes at birth to form the **ligamentum arteriosum**. The closure has long been known to be related to the higher oxygen tension in the blood following initial breathing. It is now believed to be mediated by **bradykinin**, a substance released by the lungs whose action is dependent on higher blood oxygen content.

The umbilical veins also close, becoming the ligamentum teres and the ligamentum venosum. The proximal segments of the umbilical arteries form the superior vesical arteries; the remainder forms the lateral umbilical ligaments.

IV. THE CIRCULATORY SYSTEM OF CHICK AND PIG EMBRYOS

A. The heart and the arterial system
1. 33-HOUR CHICK EMBRYO

Review the structure of the heart on the whole mount. What vessels enter the posterior end of the heart? What vessel emerges from the anterior end?

On the cross sections, find the **vitelline veins** converging into the sinus venosus. Identify the **ventral aorta** and the **first aortic arch**. Follow the **dorsal aortae** posteriorly. Is there a single dorsal aorta present at any point? How do the aortae end?

2. 48-HOUR CHICK EMBRYO
On the whole mount identify the **sinus venosus**, **ventricle, atrium, bulbus arteriosus**, and **aortic arches**. Remember that the 48-hour embryo is quite thick, and it is necessary to focus on several levels to make out the shape of the entire heart.

In the cross sections find the mandibular arches again and identify the **first aortic arches** within them. Trace these back and note that they arise from a common aortic sac, from which the **second aortic arches** also originate. The aortic sac is really the ventral aorta, which is very short in the chick embryo and emerges directly from the **bulbus arteriosus**.

Tracing back, follow the bulbus into the **ventricle**, the large chamber lying across the entire width of the body. On the side opposite the bulbus the ventricle is continuous with the **atrium**. Note that the **endocardium** is much farther away from the **myocardium** in the ventricle than in the atrium; this difference presages the future difference in thickness of the muscular walls of the two chambers. What is the functional reason for this difference in thickness between the atrium and ventricle?

Behind the atrium is a median chamber, which is attached below the esophagus by means of a **mesocardium**. This chamber is the **sinus venosus**. What are the two large vessels that enter the sinus from each side of the body wall? Follow the sinus back until it breaks down into the two **vitelline veins**.

Now return to the aortic sac and trace one set of aortic arches until they enter the **dorsal aortae**. Follow the aortae back from the point of entrance of the aortic arches and find where the two vessels come together to form the single median aorta. Farther posterior, however, the vessel is still double. Find the place where the two **vitelline arteries** spring from the aortae and spread out over the yolk sac. How far back do the aortae extend beyond this point?

3. 72-HOUR CHICK EMBRYO
Examine the surface of the heart on the whole mount. Identify the aortic arches, vitelline vessels, and all the chambers of the heart. To aid you in your studies, frequently refer to the two scanning electron micrographs of wax casts of the chick circulatory system (Figure 8.6) and to the supplement to Chapter 4 (Figure 4.52 and 4.53).

On the sections, find the upper end of the **bulbus arteriosus** and trace through the entire heart. You will find that the bulbus runs along-side the **atrium** for a considerable distance. Note that the **sinus venosus** (which may still be identified by the common cardinal veins entering it) is now near the anterior end of the heart and has become asymmetrical; on which side of the body does it now lie? A little farther back the bulbus and atrium are confluent with the thick-walled ventricle, which is as wide as the body. Dorsal to the ventricle is the **meatus venosus**, formed from the fused vitelline veins. The meatus venosus is surrounded by a mass of tissue that has already budded off the liver diverticula.

4. 10-MM PIG EMBRYO
At the end of the pharynx find the **bulbus** and

trace it back. You will note that its channel has now been split into two parts. The one on the apparent left is the **systemic aorta**, the other is the **pulmonary trunk**. The atrium has expanded so much that it now lies on both sides of the bulbus. Tracing back you will find that the two parts of the atrium come together, and the bulbus passes into the ventricle.

The atrium has become subdivided by a partition (Figure 10.10). This partition, the **interatrial septum**, is incomplete, so that the two chambers actually communicate with each through an opening called the **foramen ovale** (Figure 8.4). The ventricle, which by now has a heavy muscular wall, is similarly partially subdivided by a massive **interventricular septum**. Identify the thick **endocardial cushion** which separates the atria from the ventricles and narrows the **atrioventricular canal**. Find the much reduced sinus venosus entering the right atrium (Figure 10.16).

B. The venous system
1. 48-HOUR CHICK EMBRYO
Veins in the body wall at this stage are the **anterior and posterior cardinals**. Trace both of these from the **common cardinal veins**. The splanchnic vessels are represented by the **meatus** (or **ductus**) **venosus** formed by the fusion of the intraembryonic parts of the two **vitelline veins**.

2. 72-HOUR CHICK EMBRYO
The cardinal veins are present as in the 48-hour embryo. The posterior cardinal persists as long as the mesonephros remains functional. A new vessel, the subcardinal vein, appears under each mesonephros.

3. 10-MM PIG EMBRYO
Find the **common cardinal veins** and determine where *each* of them enters the heart. Study carefully the effect of the asymmetry of the sinus venosus on the course of these vessels. The anterior cardinals, which remain symmetrical, may now be called the **internal jugular veins** (Figure 10.3). They constitute irregular sinus-like spaces which may be readily traced into the head.

The **posterior cardinals** run back along the dor-

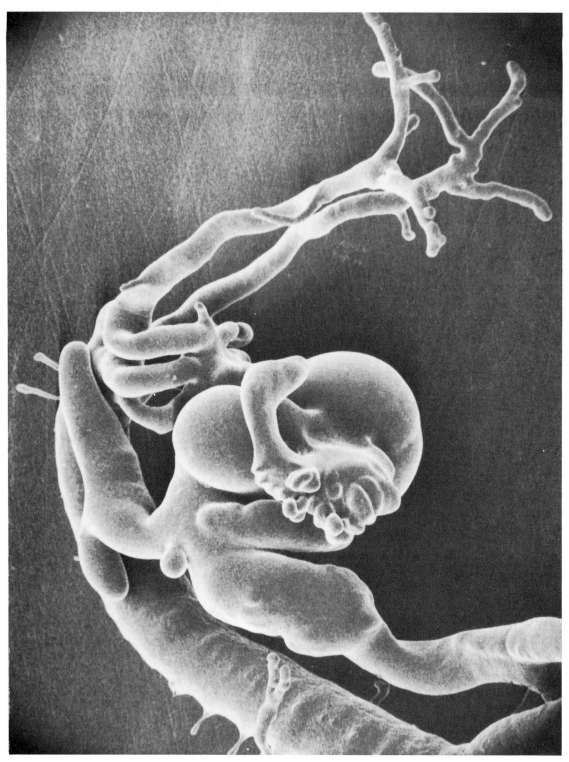

Figure 8.6A

Figure 8.6. *The circulatory system of a 96-hour chick embryo.* A. SEM B. Key to figure.

solateral corners of the mesonephros. Midway along the length of the mesonephros, the posterior cardinals are reduced or absent; and the dominant venous vessels are the **subcardinals**, which now form an extensive sinus in the kidneys.[3] More posteriorly, however, they persist.

In the umbilical cord find the two **umbilical**

3. This is the renal portal system of lower vertebrates.

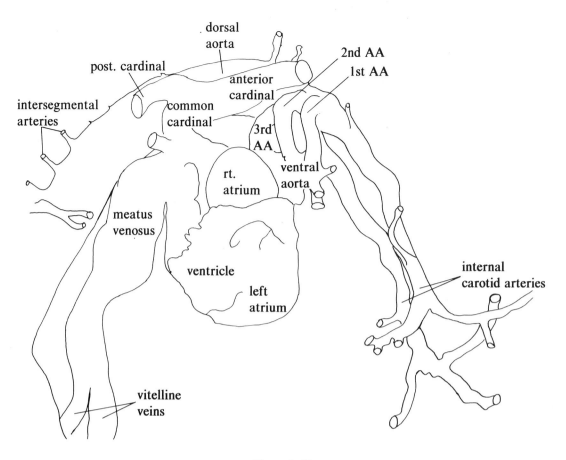

dorsal
aorta

post. cardinal

2nd AA

1st AA

anterior
cardinal

intersegmental
arteries

common
cardinal

3rd
AA

rt.
atrium

ventral
aorta

meatus
venosus

internal
carotid arteries

ventricle

left
atrium

vitelline
veins

Figure 8.6B

veins. The right one, which disappears early, is already smaller than the left. Both veins enter the liver, the left excavating in the liver substance a large channel, the **ductus venosus**, which makes its way to the **vena cava.** The presence of the ductus venosus makes it possible for fresh blood coming from the placenta to avoid circulating through the liver sinusoids. The ductus thus has the same function in the mammal as the meatus venosus does in the chick embryo, but note that they are not homologous.

One more important vein that has appeared at this stage is the **hepatic portal.** It is best identified at the level of the pancreas, where it constitutes a prominent opening in the mesentery (Figure 10.17). Follow it forward to the liver. Where does its entrance into the liver occur, relative to the entrance of the postcava? Now trace the portal back from the pancreas, noting how it weaves from one side of the gut to the other. This peculiar course is explained by the fact that the vessel is made up of parts of both vitelline veins.

Return to the pulmonary trunk in the bulbus and follow it forward. It splits into two branches, the **sixth aortic arches**, which run up to the dorsal aortae. Near its base each of these arches gives off a **pulmonary artery** to the lung. The origin of the pulmonary arteries is most easily found by looking for a pair of tiny vessels in the mesenchyme below the gut, a little behind the arches. These vessels, which are the pulmonary arteries, may then be traced forward into the sixth arches (Figure 10.10). Note that the right sixth arch is quite small; it will soon disappear. But the left sixth arch persists throughout the embryonic period as the **ductus arteriosus.**

Find the point of union of the two dorsal aortae. Note the dorsal and lateral branches which the aorta gives off. Follow one lateral branch to its destination in a glomerulus in the kidney (Figure 10.18).

At the level of the hind limb bud, find the two **umbilical (allantoic) arteries** diverging from the aorta and running forward into the umbilical cord (Figure 10.21). What is the structure that lies between these two arteries in the cord? (Figure 10.16)

Figure 8.7 *Scanning electron micrographs of chick embryos.* These embryos were bisected in the sagittal plane to reveal the sturctures visible from the cut surface of the right half of the embryo. Figure A (top figure) shows a sagittal section of a 33-hour chick embryo. Visible are the major divisions of the brain [prosencephalon (pros) with the opening of the right optic vesicle, mesencephalon (mes), and rhombencephalon (rhom)], notochord (n), foregut (f), anterior intestinal portal (aip), left vitelline vein (vv), heart (h), and ventral aorta (va). Figure B (bottom figure) shows a sagittal section of a 72-hour chick embryo. The prosencephalon is subdivided into the telencephalon (tel) and diencephalon (dien), from which arises the stalk of the optic visicle (o). Most of the mesencephalon and the metencephalon are out of view at the top of the picture. The myelencephalon can be identified by the presence of neuromeres (neu) in its ventral wall. In this picture, the digestive system is represented by the stomodeum (s) with Rathke's pouch (r) as a dorsal diverticulum and the pharynx (p) with pharyngeal pouches 1-4. The separation between pharynx and stomodeum is demarcated by remnants of the pharyngeal plate (pp). The circulatory system is represented by the ductus venosus (dv), which has been invaded by cords of liver tissue, the sinus venosus (sv), the atrium (at), the ventricle (v), the conus (c), the base of aortic arches 3 and 4 (aa), and the dorsal aorta (da). An idea of the growth of the embryo between 33 and 72 hours can be had by comparing the two pictures; the final magnification is identical, X 100.

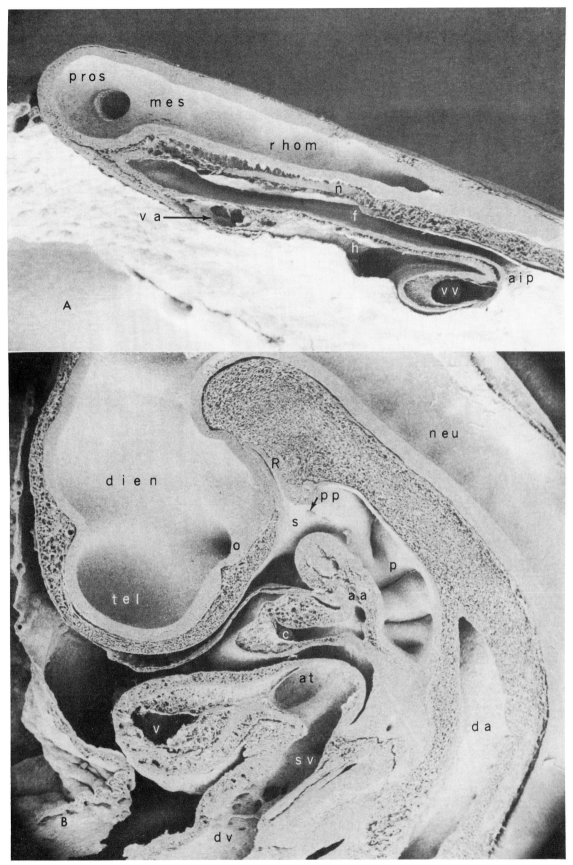

Figure 8.7

THE NERVOUS SYSTEM

I. INTRODUCTION

A. The adult nervous system

The functions of the nervous system include communication between the organism and the outside world and coordination of the function of the organs of the body. The principal cell type of the nervous system is the **neuron**. Neurons characteristically have a cell body and one or several elongated cellular extensions, the **axons** and the **dendrites**. In the adult, single axons may be very long: up to several meters in large vertebrates such as humans and elephants.

The nervous system may be divided into two broad subdivisions: the **central nervous system** and the **peripheral nervous system**. The brain and spinal cord constitute the central nervous system. The sensory and motor nerves are part of the peripheral nervous system. Information about the external world is received by sensory structures such as eyes, ears, olfactory epithelium, taste buds, and cutaneous touch receptors. The information received is conveyed to the central nervous system by way of the axons of sensory nerves. Motor nerves convey instructions from brain and spinal cord to organs in the periphery.

The cell bodies of the neurons of the peripheral nervous system may lie within the brain or spinal cord or may lie in clusters called **ganglia** outside of the central nervous system. The ganglia associated with the brain are the **cranial ganglia** (Table 9.1); those associated with the spinal cord are the **spinal ganglia**. In addition, **motor ganglia** are present in the viscera, at sites distant from the brain or spinal cord. In this exercise, you will have an opportunity to study the major divisions of the brain and spinal cord and to locate and observe the major ganglia and nerve tracts that constitute the peripheral nervous system.

B. Origin of the nervous system

The nervous system is of ectodermal origin, receiving contributions from all three major embryonic regions of the ectoderm: **neural ectoderm, neural crest and epidermis**. These three regions first become distinguishable following neurulation when these three classes of ectoderm separate from each other. Neural ectoderm becomes established as the hollow dorsally-situated neural tube from which the entire central nervous system (brain and spinal chord) will develop. With neural closure the surface of the embryo is left covered solely by epidermal ectoderm. Finally, cells of the neural crest detach from the dorsal surface of the neural tube and then disperse to distant sites in the body. The neural crest contributes cells to several of the cranial ganglia and to all of the spinal and autonomic ganglia. Epidermis and neural ectoderm also contribute to some of the cranial ganglia. A more thorough discussion of the organiza-

109

tion and development of spinal and cranial ganglia will be reserved for later sections of the chapter.

C. Organogenesis of the nervous system
1. CENTRAL NERVOUS SYTEM
Even as early as the neural plate stage, regional specialization of neural ectoderm is apparent in the amphibia with the broad anterior future brain region distinguishable from the narrow posterior region destined to form the spinal cord (Figures 3.16 to 3.18). Following neural closure, the brain becomes subdivided along the anterior-posterior axis of the embryo into **forebrain (prosencephalon)**, **midbrain (mesencephalon)**, and **hindbrain (rhombencephalon)**. This stage of development is illustrated by the 33-hour chick embryo (Figure 4.12 and 4.23). Subsequently, the prosencephalon subdivides into the anterior **telencephalon** and posterior **diencephalon** and the rhombencephalon subdivides into the anterior **metencephalon** and posterior **myelencephalon** (Table 4.1). The segments of the brain are delineated by constriction of the wall. The five major divisions are clearly seen in the 72-hour chick embryo (Figure 4.39). The telencephalon gives rise to the cerebral hemispheres; the diencephalon to a number of derivatives including the optic vesicles, hypothalamus and the posterior lobe of the pituitary (Table 4.1). The mesencephalon develops into the optic tectum. The anterior part of the optic tectum receives visual information via the optic nerve. The metencephalon develops into the cerebellum and the myelencephalon into the medulla. The lumen of the neural tube is present in the adult, both in brain and spinal cord.

Table 9.1. The Cranial Nerves.[a]

Nerve	Peripheral Structure Innervated	Origin	Function
I. Olfactory	Olfactory epithelium	Epidermis	Sensory
II. Optic	Retina	Neural ectoderm	Sensory
III. Oculomotor	Muscles of eye	Neural ectoderm	Motor
IV. Trochlear	Muscles of eye	Neural ectoderm	Motor
V. Trigeminal	Tactile receptors of face (sensory) + muscles of jaw (motor)	Neural crest (sensory) + neural ectoderm (motor)	Both
VI. Abducens	Same as V	Neural ectoderm	Motor
VII. Facial	Taste buds (sensory) + face muscles (motor)	Neural ectoderm	Both
VIII. Acoustic	Inner ear	Neural crest	Sensory
IX. Glossopharyngeal	Tongue & pharynx (sensory) + muscles of pharynx (motor)	Neural crest (sensory) + neural ectoderm (motor)	Both
X. Vagus	Skin around ear & visceral organs (larynx, trachea, heart, lungs, gut) (sensory) + viscera (motor)	Same as X	Both
XI. Spinal accessory	Neck & back muscles	Neural ectoderm	Motor
XII. Hypoglossal	Muscles of tongue	Neural ectoderm	Motor

[a]An easy way to remember the twelve cranial nerves is to take the first letter of each word in the following: "On old Olympus towering top, a Finn and German viewed some hop." If you also want to remember their functions, try this: "Some say marry money but my brother says bad business marry money."

In the brain local enlargements of the neural canal are known as **ventricles**.

2. SENSE ORGANS

Sensory tissues are also derived from ectoderm: the nasal sensory epithelium, the inner ear, and cutaneous receptors from the epidermis, and the rods and cones of the eye from neural ectoderm. The eye is a compound organ with the lens and cornea derived from epidermis and the neural retina (the parent tissue for the rods and cones and the cells of the optic nerve) and pigmented retina (a single layer of pigment-containing cells that lie behind the neural retina) derived from neural ectoderm (i.e., the optic vesicle; Figure 4.31). A characteristic series of events in the development of components of sense organs of epidermal origin (e.g., nasal epithelium, lens, inner ear) is the establishment of **placodes**, which are local thickenings of epidermis. Following epidermal thickening, the placode becomes a cup-shaped depression. In the case of the nasal placode (Figures 4.47; 10.6 through 10.8) the tissue of the depression differentiates into the sensory lining of the nose and projects nerve axons back to synapse with the brain. Once functional, these axons will convey olfactory information to the central nervous system. The olfactory nerve is the first cranial nerve (Table 9.1). The lens and auditory placodes continue the inpocketing process to the extent that internal spheres of epidermal tissue are established which differentiate ultimately into lens or inner ear (lens: Figures 4.30, 4.31, 10.1, 10.4; otic vesicle: Figures 4.29, 4.30, 10.1).

3. THE PERIPHERAL NERVOUS SYSTEM

The peripheral nervous system associated with the brain is represented by the cranial nerves and their associated ganglia. Some of the cranial nerves are purely sensory in function, some purely motor, and some are a mixture of sensory and motor fibers (Table 9.1).

The olfactory nerve originates from the sensory epithelium of the nose, which, of course, originates from nasal placode epidermis. The optic nerve arises from the neural retina. Retinal ganglion cells project axons back to the brain, which synapse at the optic tectum (derived from the mesencephalon). The acoustic ganglion develops as an aggregate of neural crest cells associated with the otocyst. The motor components of the cranial nerves have their cell bodies situated in the brain with the axons passing into the periphery. The sensory components of the mixed-function cranial nerves (V, VII, IX, X, XI) are of neural crest origin and arise from ganglia situated distant from the brain. Table 9.1 summarizes functional and developmental information on the twelve cranial nerves.

In the trunk, the spinal cord receives sensory input over the dorsal root fibers whose cells bodies are located in the segmentally arranged dorsal root ganglia that develop in the anterior half of each somite (Figure 10.14). The nerves convey sensory input from receptors in the skin to the spinal cord. Spinal motor nerves are of two types: voluntary (which innervate skeletal muscle) and involuntary (which innervate smooth muscle and glands). The voluntary motor nerves have their cell bodies situated in the floor of the spinal cord with the axons passing from the ventral corners of the spinal cord and synapsing at their distal ends with skeletal muscle fibers. The involuntary pathways involve two nerves in series: the involuntary preganglionic axons whose cell bodies lie in the spinal chord and the involuntary postganglionic fibers whose cell bodies lie in ganglia distributed in the viscera. The preganglionic axons synapse with postganglionic neurons in the ganglia and the latter synapse with the end organs (smooth muscle or gland). The spinal sensory nerves and involuntary postganglionic nerves are of neural crest origin and the voluntary and involuntary preganglionic motor nerves are of neural ectoderm origin.

II. THE NERVOUS SYSTEM IN CHICK AND PIG EMBRYOS

A. 33-hour chick embryo

Review the general organization of brain and spinal cord. The main features can best be appreciated in the whole mount. The anterior two thirds of the neural plate has rolled up to form the brain and the anterior portion of the spinal cord (Figures 4.5, 4.7, 4.8). Posteriorly, the neural epithelium is represented by the trough-shaped neural plate. The three major **divisions** of the brain, common to all vertebrates, are evident: forebrain (**prosencephalon**), midbrain (**mesencephalon**), and hindbrain (**rhombencephalon**; Figure 4.12).

B. 48-hour chick embryo

Review the organization of the brain and sense organs. The three primary brain vesicles have now further subdivided to give rise to the five brain vesicles (Table 4.1). Use both whole mount and cross sections in your study.

C. 72-hour chick embryo

1. BRAIN

Review the organization of the brain in the whole mount and cross sections. The subdivision of the forebrain (prosencephalon) into **telencephalon** and **diencephalon** and the hindbrain (rhombencephalon) into **metencephalon** and **myelencephalon** is now obvious, especially in the whole mount.

2. SENSE ORGANS

a. Nasal placodes.

The nasal placodes arise as thickenings of the ectoderm in the ventrolateral region of the head in the region of the telencephalon (Figure 4.47). In cross sections of 72-hour embryos, the placodes form shallow cup-like structures in the ectoderm. The nasal placode will eventually develop into the olfactory epithelium. The cells of this epithelium develop into chemoreceptors that line the nasal passage. These cells also send axons into the telencephalon. This bundle of axons eventually becomes the first cranial (olfactory) nerve.

b. Eye.

The eye is a compound structure, with the retina originating from the brain (diencephalon) and the lens and cornea from epidermal ectoderm in the head region. The retinal portion consists of the thick pseudostratified **neural retina** and the single-layered **pigmented retina** (Figure 4.46). As the eye develops beyond the stage seen here, the former layer gives rise to the photoreceptor cells (rods and cones), and a variety of neural cells, including the retinal ganglion cells that form the optic nerve (cranial nerve II). The lens is by this stage a closed vesicle, no longer connected to head epidermis (Figure 4.46).

c. Otocyst.

Another sensory structure derived from epidermal ectoderm in the head region, the otocyst forms the primordium for the lining of the inner ear, including the sensory epithelium.

3. CRANIAL NERVES

Not all cranial nerves are well-developed by 72 hours, but portions of some cranial nerves can be identified.

a. CNIII (Oculomotor)

The nerve arises from the floor of the mesencephalon as two short sprouts that are directed toward the stomodeum.

b. CNVIII (Acoustic ganglion of auditory nerve) and CNVII (Geniculate ganglion of facial nerve)

At this stage, these two nerves are associated with each other and form a common bundle directly anterior to the otocyst (Figure 4.40). Separation into acoustic and geniculate masses will occur later in development. Trace the nerve deeper into the embryo until the facial nerve becomes distinct as a small darkly-staining mass anterior to the main ganglionic mass. The nerve passes into the second pharyngeal arch.

c. CNV (Semilunar ganglion of the trigeminal nerve)

This is easily distinguished as a large darkly-staining region closely associated with the lateral wall of the myelencephalon (Figure 4.40). Follow the nerve back in your set of cross sections: as one moves posteriorly it can be seen that the ganglion becomes divided into three sections.

d. CNIX (Superior ganglion of the glossopharyngeal nerve)

This ganglion is located just caudal to the otocysts. In Figure 4.40 it can be seen as a small dark mass (unlabeled) posterior to the otocyst. Using the compound microscope, trace the glossopharyngeal nerve deeper into the embryo to the level of the pharynx. Here the nerve enlarges again as the **petrosal ganglion** and fuses with an ectodermal placode just posterior to the third pharyngeal pouch.

e. CNX, CNXI, CNXII
These nerves are rather difficult to locate in many specimens and can be neglected for the present.

4. SPINAL CORD
The different regions of the spinal cord and the spinal sensory and motor nerves will be examined in the 10 mm pig.

D. 10-mm pig embryo
1. BRAIN
The paired cerebral hemispheres of the telenchaphalon are easily seen in cross sections (Figure 10.8). The nasal placodes are often useful markers for the telencephalon. The diencephalon can be traced anteriorly and easily identified by a variety of diagnostic structures, such as the optic stalks and eye cups (Figure 10.4), and Rathke's pouch (Figure 10.3).

The mesencephalon is recognizable by its almost circular shape in cross section (Figure 10.1). Finally, the myelencephalon, with its prominent neuromeres and thin roof, is also closely associated with the otic vesicles (Figure 10.1).

2. SENSORY ORGANS
Locate infolded **nasal placodes** (Figures 10.6 through 10.8), **eyes** (Figures 10.3, 10.4) (distinguish between **neural retinal** and **pigmented retina**), and **otocysts** (Figure 10.1). The **endolymphatic duct** can be seen in the latter as a small sprout directed toward the brain.

3. CRANIAL NERVES
a. Nerves associated with telencephalon:
CNI (**Olfactory nerve**). Originates from nasal placode. Not always very distinct. Use 10X objective to locate.

b. Nerves associated with diencephalon:
CNII (**Optic nerve**). Originates in the neural retina to run along the underside of the optic stalk. Sometimes difficult to find at this stage. Use 10X objective to locate.

c. Nerves associated with mesencephalon:
CNIII (**Oculomotor nerve**). Arises from floor of mesencephalon and can be traced backward toward the eye. The nerve eventually innervates several eye muscles.

d. Nerves associated with the lateral walls of the myelencephalon, anterior to the otocyst.
1) CNV (Semilunar ganglion of the **trigeminal nerve**). The large and very conspicuous ganglion lateral to the widest (anterior) region of the myelencephalon (Figure 10.1). The nerve divides into three branches (hence the name: trigeminal). Of these, the most prominent is the **mandibular ramus** which arises from the inner margin of the ganglion and passes into the mandibular process of the 1st pharyngeal arch (it is the unlabeled light circle in the mandibular processes of Figure 10.3).

2) CNVII (Geniculate ganglion of **facial nerve**) and CNVIII (Acoustic ganglion and **auditory nerve**). The two ganglia are closely associated with each other, but by this stage they can be distinguished (Figure 10.1). Ganglion VIII is in direct contact with the otocyst and VII is located in front of this. The facial nerve inserts into the myelencephalon and the distal portion of the nerve innervates the side of the head and passes into pharyngeal arch II. The auditory nerve innervates the otocyst and will eventually transport stimuli from the sensory epithelium of the ear to the brain. Trace its nerve fibers into the myelencephalon.

e. Nerves associated with the lateral walls of the myelencephalon, posterior to otocyst
1) CNIX (**Glossopharyngeal nerve**). The nerve enters the brain just inside the posterior margin of the otocyst and passes deep into the tissues of the head to innervate the tongue and pharynx. The ganglion situated beside the otocyst is the **superior ganglion**. Trace the nerve into pharyngeal arch III to locate a second ganglion, the **petrosal ganglion**.

2) CNX (**Vagus nerve**). Located just behind the superior ganglion. Find the **jugular ganglion** of nerve X (Figure 10.1). Trace the nerve deeper into the head to locate the **nodose gan-**

113

glion of nerve X (dark mass a little behind pouch III in Figure 10.4).

3) CNXI (**Spinal accessory nerve**). Locate the origin of this nerve as a series of small roots rather high up on the lateral wall of the myelencephalon posterior to the jugular ganglion. Two branches of the nerve can be located: the internal branch that parallels the vagus and the external branch that runs ventrally along the spinal cord.

f. Nerves associated with the floor of the myelencephalon

1) CNXI (**Abducens**). This small nerve can be located most easily in sections that just graze the floor of the myelencephalon. The nerve originates adjacent to the anterior margins of the geniculate ganglion and passes forward toward the eyes.

2) CNXII (**Froriceps ganglion of the hypoglossal nerve**). Easily located as a series of dots bridging the gap between posterior end of the myelencephalon and the spinal cord in sections where the two structures are separate.

4. SPINAL CORD
Examine a section through the trunk with the compound microscope (Figure 10.12). Identify the following:

a. Spinal cord

b. Neurocoel
Central canal in spinal cord.

c. Ependymal layer
Epithelium lining the neurocoel. Contains supporting ependymal cells and mitotic neural and glial cells that migrate temporarily from the mantle layer into the ependymal layer during mitosis and return to the mantle layer at is completion.

d. Mantle layer
Intermediate layer with cell bodies of neurons and glial cells. Neuroblasts concerned with reception of sensory messages for dorsal root (sensory) neurons are situated dorsally in the mantle layer. Cell bodies of motor neuroblasts are situated in lateral and ventral regions of the mantle layer.

e. Marginal layer
Superficial layer of lightly staining tissue, composed of the axons of the spinal cord neurons. In the dorsal portion of the superficial layer (dorsal to the site of the insertion of the dorsal root fibers), axons of the spinal ganglia course both cranial and caudal in the marginal layer. In lateral and ventral regions of the marginal layer, the axons connect adjacent regions of the spinal cord or pass to or from the brain.

5. SPINAL GANGLION
Occur as a series of condensations situated on either side of the spinal cord. Identify:

a. Dorsal root ganglia
Darkly staining masses of nerve cell bodies situated pair-wise on each side along the length of the spinal cord.

b. Dorsal roots
Sites of insertion of nerve fibers into the dorsolateral margin of the spinal cord. The nerves are spinal sensory nerves which originate from the neural crest and in the mature organism carry sensory information from the periphery into the central nervous system.

c. Ventral root
Motor nerves whose cell bodies lie in the mantle layer of the spinal cord and whose axons emerge from the ventrolateral margins of the spinal cord. These nerves originate from neural ectoderm and convey stimuli to effector organs (i.e., muscle, glands, etc.) in the periphery.

d. Spinal nerve
The nerve formed by junction of dorsal and ventral root axonal fibers.

e. Dorsal ramus
Branch that passes dorsally just beyond the junction of dorsal and ventral root fibers.

f. Ventral ramus
Branch of spinal nerve that is directed ventrally.

114

g. Brachial Plexus, Lumbar Plexus

The spinal nerve supply in the regions of the fore- and hindlimbs is augmented by a local hypertrophy of the spinal cord itself and the dorsal root ganglia. The brachial plexus is produced by nerve fibers that pass from one spinal nerve to the next in the ventral rami of spinal nerves III, IV, and V. Trace this into the forelimb bud. The lumbar plexus services the hind limb.

6. AUTONOMIC GANGLIA

The motor nerves emerging from the ventral branch include both voluntary and involuntary motor nerves. The former innervate voluntary (skeletal) muscle while the latter innervate involuntary (smooth) muscle and a variety of glands in the viscera and skin. Voluntary nerve fibers pass directly to the target organs but the involuntary fibers originating in the spinal cord synapse with a second set of neurons that then make the terminal connections with the end organs. The neurons originating in the spinal cord are of neural ectodermal origin and are known as the involuntary preganglionic nerves whereas the neurons that make connections with the end organs are known as the involuntary postganglionic nerves and these are of neural crest origin. Synapse is made at the autonomic ganglia. The cell bodies of the postganglionic neurons are located here. Automonic ganglia appear as darkly staining masses situated near visceral sites. They can easily be located in sections passing through the forelimbs as darkly staining masses in the mesenchyme above the dorsal aortae.

PICTURE ATLAS OF THE 10-mm PIG EMBRYO

The next eleven plates include representative cross sections from the 10-mm pig embryo. These photographs should prove a useful reference in your studies of organogenesis.

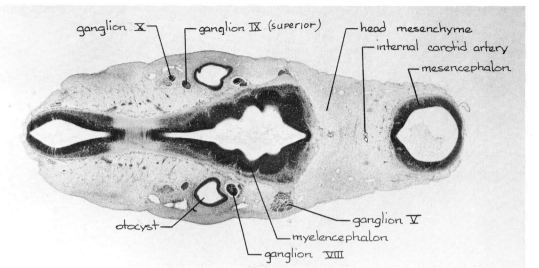

chapter 10

Figure 10.1

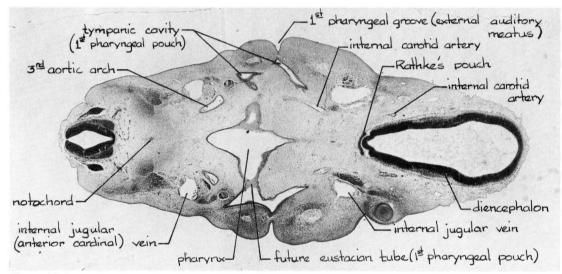

Figure 10.2

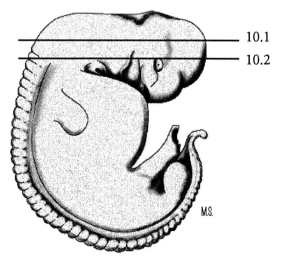

10.1
10.2

M.S.

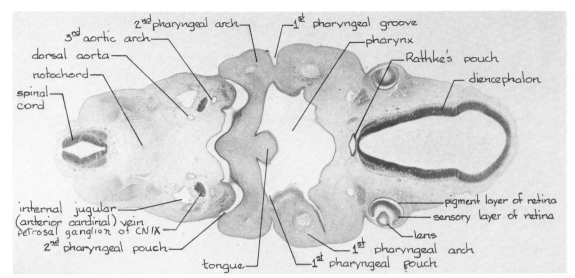

Figure 10.3

Labels (Figure 10.3):
3rd aortic arch
dorsal aorta
notochord
spinal cord
internal jugular (anterior cardinal) vein
petrosal ganglion of CN IX
2nd pharyngeal pouch
2nd pharyngeal arch
1st pharyngeal groove
pharynx
Rathke's pouch
diencephalon
pigment layer of retina
sensory layer of retina
lens
1st pharyngeal arch
1st pharyngeal pouch
tongue

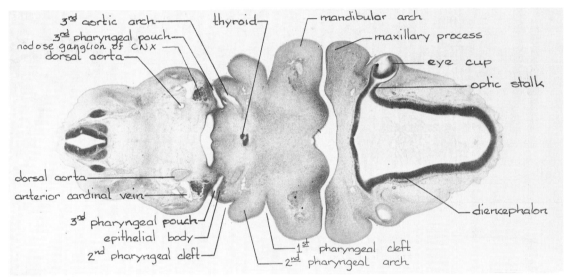

Figure 10.4

Labels (Figure 10.4):
3rd aortic arch
3rd pharyngeal pouch
nodose ganglion of CN X
dorsal aorta
thyroid
mandibular arch
maxillary process
eye cup
optic stalk
dorsal aorta
anterior cardinal vein
3rd pharyngeal pouch
epithelial body
2nd pharyngeal cleft
1st pharyngeal cleft
2nd pharyngeal arch
diencephalon

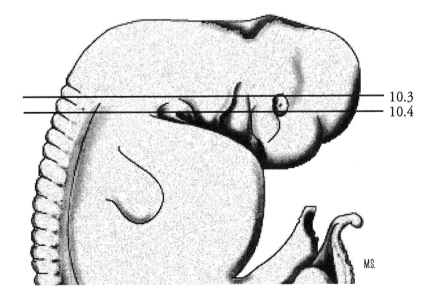

10.3
10.4

M.S.

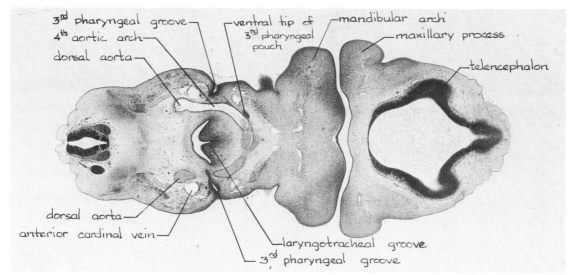

Figure 10.5

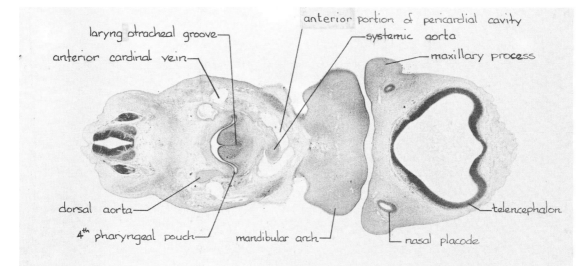

Figure 10.6

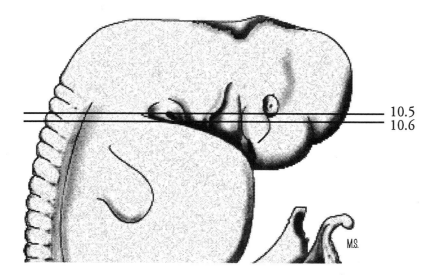

120

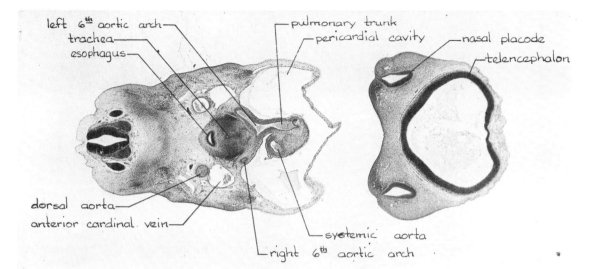

Figure 10.7

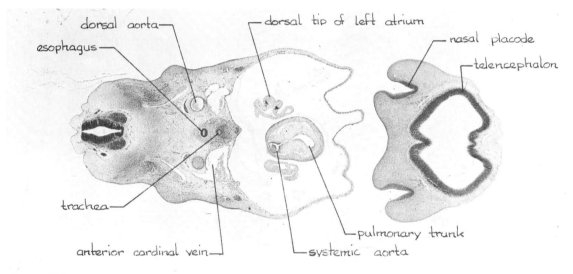

Figure 10.8

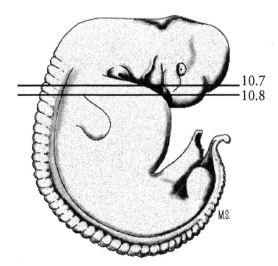

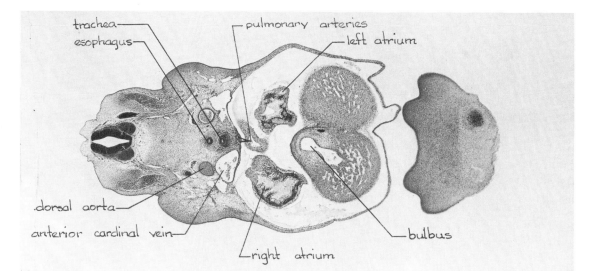

Figure 10.9

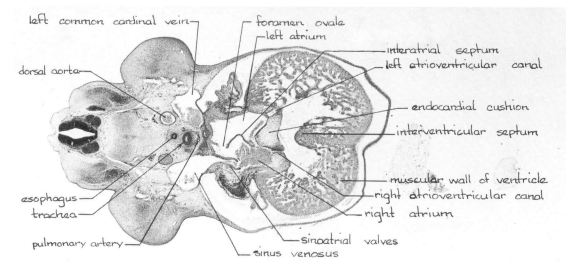

Figure 10.10

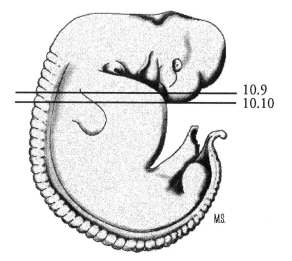

122

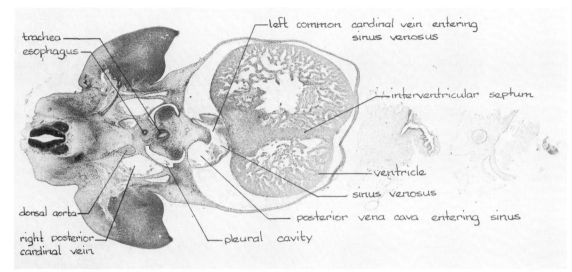

Figure 10.11

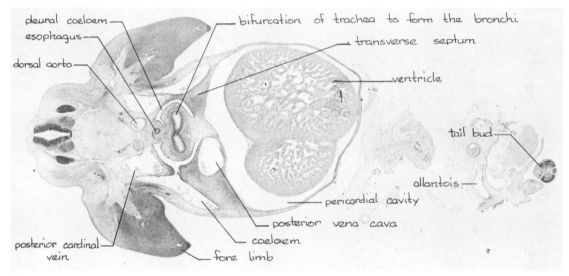

Figure 10.12

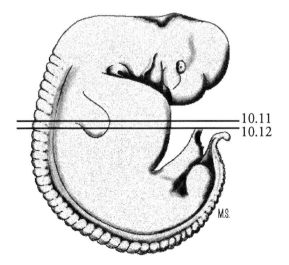

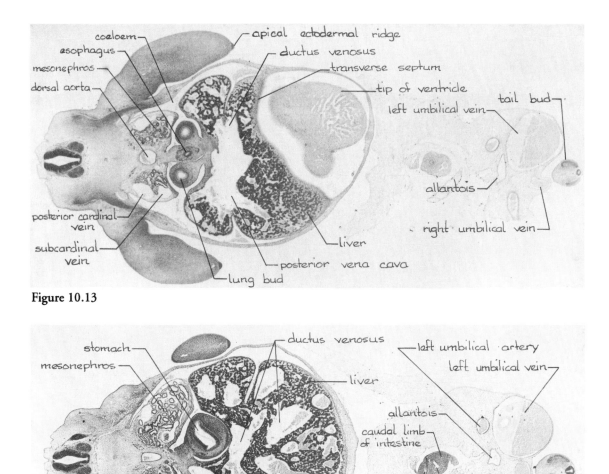

coeloem
esophagus
mesonephros
dorsal aorta

apical ectodermal ridge
ductus venosus
transverse septum
tip of ventricle
left umbilical vein
tail bud

allantois
right umbilical vein

posterior cardinal vein
subcardinal vein
lung bud
posterior vena cava
liver

Figure 10.13

stomach
mesonephros
ductus venosus
liver
left umbilical artery
left umbilical vein
allantois
caudal limb of intestine
cranial limb of intestine
superior mesenteric artery
right umbilical vein

posterior cardinal vein
subcardinal vein
posterior vena cava

Figure 10.14

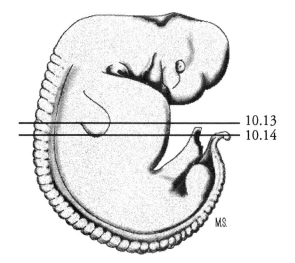

10.13
10.14

M.S.

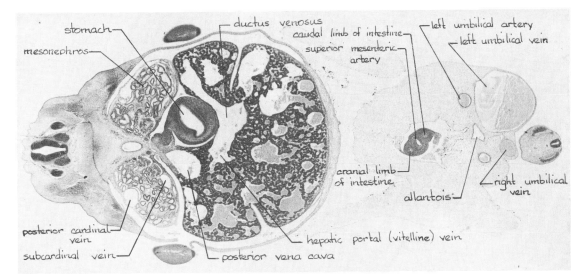

Figure 10.15

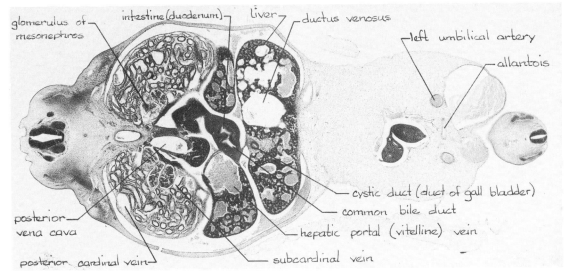

Figure 10.16

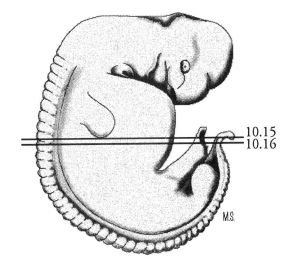

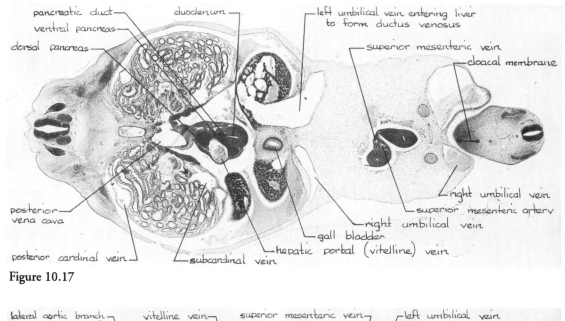

Figure 10.17

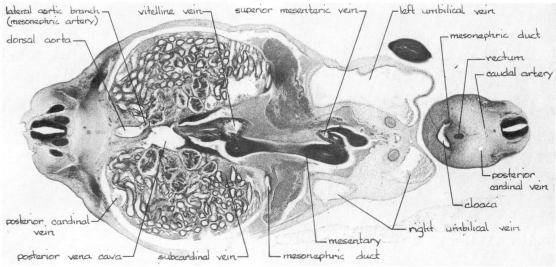

Figure 10.18

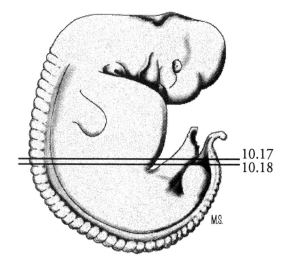

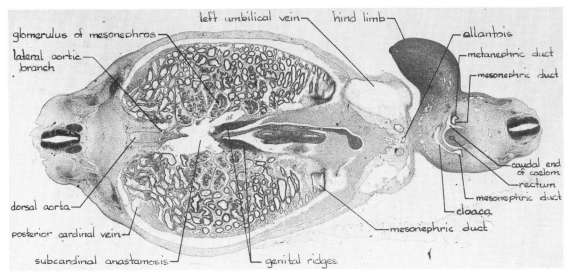

Figure 10.19

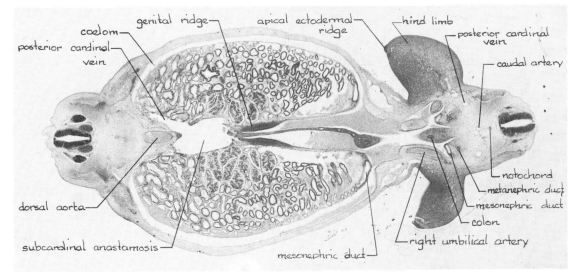

Figure 10.20

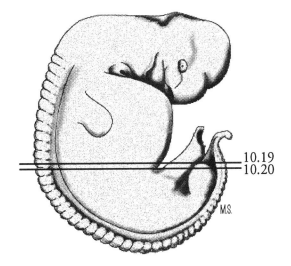

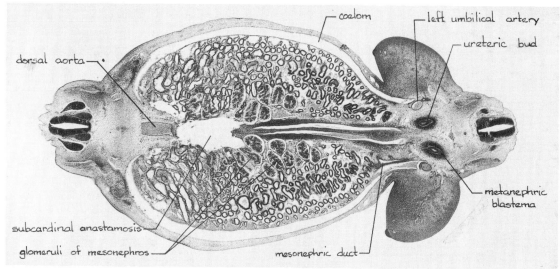

Figure 10.21

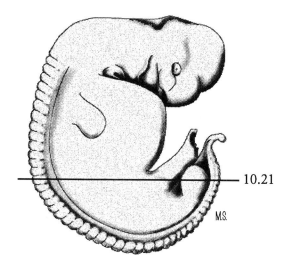

10.21

M.S.

A23187
> An ionophore that selectively carries calcium across cell membranes.

acrosome
> A specialized, membrane-bound vesicle in the anterior portion of the sperm head containing hyaluronidase and other macromolecules that function during sperm binding and fertilization.

acrosome reaction
> The name given to the exocytotic event when the acrosomal membrane fuses with the sperm cell membrane, releasing the contents of the acrosome.

albumen
> The white of a bird's egg. Provides protein to the growing embryo, helps prevent dehydration, and contains anti-bacterial compounds.

albumin
> The major protein of serum. Produced in the liver, it contributes to the osmotic pressure which keeps water in the vascular system, and serves as a carrier protein for many molecules being transported in the blood.

alecithal
> A small egg containing little yolk (e.g. echinoderms). Synonym for microlecithal.

allantoic stalk
> Membranous tube in the umbilicus connecting the allantois to the cloaca or urogenital sinus; carries urine to the allantois in mammals.

allantois
> Extraembryonic membrane formed by an outpocketing of the hindgut. Forms a storage compartment for urine. Provides the vasculature of the placenta in mammals, and serves a respiratory function in amniotes.

amnion
> Extraembryonic membrane that encloses the embryo, and is derived from the somatopleure. Ectodermal component is continuous with the embryonic ectoderm.

amniote
> Higher vertebrates (birds, reptiles, and mammals) that possess an amnion during development.

amniotic fold
> Formed as the somatopleure expands and folds around the embryo, first arising anterior to the head and then developing along the sides and finally around the tail. The folds eventually meet, fuse, and separate to form the amnion and chorion.

anamniote
> Animals that do not possess an amnion anytime during development.

animal pole
> Center of the animal hemisphere; the portion of the anamniote egg that contains the pronucleus, the least yolk, and the most pigment.

anterior cardinal veins

Paired veins that drain the head of the early embryo and empty into the common cardinal veins and then into the sinus venosus. Give rise to the internal jugular veins, the superior vena cava, and the veins of the cerebrum.

anterior intestinal portal

The point at which the foregut opens into the midgut, which lies open to the yolk sac. Moves posteriorly as the lateral body folds meet and fuse at the ventral surface, eventually meeting the posterior intestinal portal to form the enclosed gut and the yolk stalk.

anteroposterior sequence

Proceeding from the head toward the tail, so that the anterior segments of the embryo are more developed than the posterior.

anus

Posterior opening of the gut.

aortic arches

Paired connections between the dorsal and ventral aortae found in the mesenchyme of the pharyngeal arches. There are six pairs: the first, second and fifth degenerate, the third persists as part of the carotid arteries, the fourth forms the arch of the aorta and the sixth forms the ductus arteriosus and part of the pulmonary vasculature.

aortic trunk

The large artery that carries blood from the left ventricle to the systemic circulation. Develops by the partitioning of the bulbus cordis into two tubes.

archenteron

The gut primordium formed during gastrulation. Its epithelial lining is endoderm.

area opaca

Area of the blastoderm that surrounds the area pellucida in the chick and lies directly on the yolk. Contains blood islands and the developing vitelline vessels.

area pellucida

Transparent area of the chick blastoderm immediately surrounding the embryo. It lies over the subgerminal cavity, which is fluid filled and contributes to the transparent appearance. The primitive streak forms in the area pellucida during gastrulation.

atrioventricular canal

Opening between the atrium and the ventricle. In the adult, flow of blood through the canal is regulated by two valves , the mitral valve (left heart) and the tricuspid valve (right heart). Failure or insufficiency of these valves causes heart murmurs, and are a major cause of heart failure in many species.

atrium

The right heart chamber, which receives blood returning from the systemic circulation in the adult, and the left heart chamber receiving blood returning from the pulmonary circulation. The blood is then passed into the ventricle, which pumps it out either to the lung (right) or to the aorta and the rest of the body (left). In the embryo there is a single atrial chamber at first, which receives blood from the sinus venosus and passes it to the ventricle. Later, as the atrium is divided, an opening remains in the septum (the foramen ovale) that bypasses the non-functioning lungs and shunts blood directly from the right to the left atria. This foramen closes at birth (or hatching); failure of closure may cause a heart murmur, or if very severe an inability to oxygenate the blood.

130

autonomic nervous system
> Division of the nervous system that innervates smooth muscle and glands. Regulates the digestive system, respiration, and cardio-vascular functions. Responsible for the "fight or flight" response, which prepares the body to respond to danger.

axon
> A nerve process extending from the nerve cell body to the target synapse or muscle endplate. Carries action potentials generated in the nerve cell body to the innervated structure. It can be many feet long in large animals.

bilateral symmetry
> An organism in which the medial sagittal plane divides it into right and left halves that are essentially mirror images of each other. Characteristic of all vertebrates.

bipinnaria larva
> Free-swimming (pelagic) larvae of the starfish comparable to the pluteus larvae of the sea urchin.

blastocoel
> Fluid-filled cavity that forms in the center of the morula. As the blastocoel expands, the morula forms a hollow ball of cells called a blastula.

blastoderm
> The flat sheet of cells lying on the surface of the yolk in the chick embryo at the blastula stage.

blastodisc
> Circular disc of yolk-poor cytoplasm that lies on the surface of the yolk in the chick egg. Develops into the embryo and extraembryonic membranes of the chick.

blastopore
> Opening from the exterior into the archenteron. Delineates the point of origin of the archenteron, and marks the posterior end of the embryo.

blastula
> Early developmental stage characterized by a hollow ball of embryonic cells surrounding a fluid-filled cavity called the blastocoel.

blood islands
> Condensations of mesodermal cells in the area opaca on the yolk sac of amniotes; earliest site of formation of red blood cells.

Bowman's capsule
> Thin-walled cellular sac that surrounds a tuft of capillaries (or glomerulus) in the kidney. Responsible for the filtration of fluid and solutes out of the blood.

Brachial plexus
> Site of interconnection between the multiple nerves that leave the spinal cord at the base of the neck and innervate the forelimbs and parts of the trunk of mammals. Damage to this plexus is a common cause of forelimb paralysis.

broad ligament
> Connective tissue "sling" composed of the mesovarium, which supports the ovary, the mesosalpinx, which supports the oviduct, and the mesometrium, which supports the uterus in the abdomen. It attaches to the body wall in the lumbar area. Inflammation of the mesosalpinx and mesovarium are common causes of infertility.

bulbus cordis

Chamber of the heart connecting the ventricle to the ventral aorta. In mammals, it splits to form the aortic and pulmonary trunks. Sometimes known as bulbus arteriosus or conus arteriosus.

bulbus arteriosus

Synonym for bulbus cordis.

capacitation

A necessary process for fertilization during which mammalian sperm acquire the ability to undergo the acrosome reaction. Occurs naturally in the oviduct in the presence of follicular fluid.

caudal

Of or relating to a posterior position.

central nervous system

Portion of the nervous system consisting of the brain and spinal cord; integrates nervous function of the whole organism.

Cephalochordata

Subphylum of the phylum Chordata; includes *Amphioxus*.

Chordata

Phylum including vertebrates and two invertebrate subphyla, Urochordata (tunicates) and Cephalochordata (*Amphioxus*). Characterized by having a dorsal hollow nerve cord, a notochord, and internal gill structures sometime during development.

chorioallantoic membrane

Extraembryonic membrane composed of the fused chorion and allantois. In mammals, this membrane comprises a major portion of the placenta. Serves as the major respiratory organ in the chick embryo.

chorion

The outermost extraembryonic membrane of the amniote embryo. Develops simultaneously with the amnion by growth of the amniotic folds in the chick; formed by trophoblast cells in mammals. Fuses with the allantois to form the chorioallantoic membrane, which is vascularized by the allantois.

cleavage

Mitotic divisions of the zygote that occur without intervening cell growth.

cloaca

The caudal chamber of the digestive tract. It receives the nephric ducts, as well as the openings of the reproductive tract and the allantois. Forms the bladder, rectum and part of the vagina in mammals. In birds, it forms the common opening for the urogenital tract and gut.

cloacal membrane

The two-layered structure that forms at the site of fusion of the cloaca with the ectoderm; opens to form the anus.

coelom

Main body cavity of the early embryo, lined by mesoderm. Eventually subdivides to form the pericardial, thoracic, and peritoneal cavities.

common cardinal veins
> Paired veins formed by the junction of the anterior and posterior cardinal veins. Empty into the sinus venosus and form part of the anterior vena cava of the adult mammal. In humans it is known as the duct of Cuvier.

common bile duct
> Duct originating from the neck of the liver diverticulum. Connects the cystic duct of the gall bladder with the hepatic duct and empties into the small intestine (duodenum).

corpora lutea
> Structure formed in the ovary from the cells of the recently ovulated follicle under the influence of luteinizing hormone. Secretes the progesterone necessary to maintain pregnancy.

cortical granules
> Protease-containing vesicles located immediately beneath the oocyte plasma membrane that are triggered to release their contents by an increase in cytoplasmic calcium at the time of sperm fusion. In the sea urchin, the proteases degrade sperm receptors on the oocyte surface and break the ties holding the vitelline membrane close to the plasma membrane, contributing to "lift-off" and the slow block to polyspermy.

cranial nerves
> Twelve paired nerves associated with the brain but part of the peripheral nervous system. They can be of sensory, motor or mixed function.

cranial ganglia
> Grouping of nerve cell bodies at the base of a cranial nerve.

cytokinesis
> The partitioning of cytoplasm into daughter cells during mitosis or meiosis.

delamination
> The process of splitting into separate cell layers that occurs during gastrulation of the chick embryo. Some cells separate from the blastoderm and form the hypoblast, leaving the epiblast dorsal to a space known as the blastocoel.

dendrite
> Process extending from a nerve cell body that carries efferent information to the cell.

dermatome
> Epithelial portion of the somite lying closest to the epidermis that forms the dermis of the skin; of mesodermal origin.

diencephalon
> Division of the brain derived from the caudal portion of the prosencephalon. The optic stalks extend from it as do the epiphysis and the infundibulum. The epithalamus, thalamus and hypothalamus are also of diencephalon origin.

diverticulum
> A blind pouch or outpocketing arising from a hollow organ.

dorsal aortae
> Paired arteries running longitudinally in the embryo that join together caudal to the pharynx to form the descending aorta.

dorsal hollow nerve tube

A characteristic of all chordates; a hollow tube of neural ectoderm located dorsally that allows integration and transmission of information from one part of the body to other parts. Forms the spinal cord (neural tube) in vertebrates.

dorsal lip

Edge of the blastopore nearest the future dorsal surface of the embryo. The first part of the blastopore to form in the amphibian. Site of Spemann's Organizer.

dorsal ramus

Dorsal branch off the spinal nerve.

dorsal root

Site where sensory nerves enter the spinal cord.

dorsal root ganglia

Nerve cell bodies of the sensory nerves located just lateral to the spinal cord. Cells derived from the neural crest.

Ducts of Cuvier

Synonym for common cardinal vein.

ductus venosus

A channel through the liver tissue that empties into the sinus venosus. Closes shortly after birth.

ductus arteriosus

Vessel connecting the pulmonary trunk and aortic trunk. Allows shunting of blood away from the lungs in the embryo. Closes shortly after birth.

ectoderm

One of the three primary germ layers formed during gastrulation. Gives rise to the epidermis of the skin, the central nervous system, and the neural crest.

endocardial cushion

Connective tissue that forms the atrioventricular valves.

endocardium

Cell layer that lines the heart lumen; derived from splanchnic mesoderm.

endoderm

Innermost of the three primary germ layers. Lines the digestive tract, respiratory tract, and gives rise to the glandular organs associated with the gut.

endolymphatic duct

Stem of the otic vesicle in the embryo; becomes part of the inner ear.

ependymal layer

Epithelial cells lining the neurocoel.

epiblast

Layers of cells lying above the blastocoel in discoid embryos.

epidermis

Cells of ectodermal origin that form the outer epithelial layer of the skin.

esophagus

Tubular organ that carries food and water from the pharynx to the crop or stomach. Formed as part of the foregut, and lined with endoderm.

external auditory meatus

Opening from the external surface of the head to the tympanic membrane (eardrum); derived from the first pharyngeal groove and is thus lined with ectoderm.

extraembryonic coelom

That part of the coelom outside the body of the embryo; the space between the amnion or yolk sac and the chorion. It is a true coelom since it is completely lined with mesoderm. In the mammal and avian embryo, the space is nearly filled by the expanding allantois at later stages.

fertilization envelope

The vitelline membrane of the egg after it has thickened and hardened as a result of the action of cortical granule contents at the time of fertilization; it results in the "slow block" to polyspermy and protects the developing embryo.

flexure

The bending of the embryo at the head, neck or tail caused by the uneven rate of tissue growth.

foramen ovale

The hole that remains open in the interatrial septum during fetal life and shunts blood directly from the right atrium to the left atrium, thereby bypassing the pulmonary circulation.

forebrain

The part of the embryonic brain also known as the prosencephalon. Later divides into the telencephalon and diencephalon.

foregut

The part of the gut that extends cranially from the anterior intestinal portal and is formed by the head fold as it grows caudally. Widens into the pharynx for part of its length. Gives rise to the pharynx, esophagus, and stomach, as well as to the majority of the respiratory tract.

gall bladder

Thin-walled sac derived from the liver diverticulum. Connected to the liver by a system of ducts and to the intestine by the cystic duct and common bile duct. Stores bile and releases it into the intestine in response to the presence of food.

ganglia

Collection of nerve cell bodies outside the central nervous system.

gastrula

Stage of embryonic development following the blastula stage; characterized by movement of cells to new positions in the embryo, producing the three primary germ layers.

genital ridge

Thickening of the splanchnic mesoderm medial to the mesonephros. Site to which the primordial germ cells migrate. Develops into the somatic tissue of the testis or ovary.

gill region
> Area of the embryo where the gut tube is expanded laterally, including the pharyngeal arches, pouches and grooves. Develops into functional gills in amphibian embryos and adult fish.

glomerulus
> Tufts of capillaries that filter the blood in mesonephric and the metanephric kidneys.

glossopharyngeal nerve
> Cranial nerve 9. Innervates the third pharyngeal arch in the embryo, and the tongue and part of the pharynx in the adult.

gray crescent
> A gray area between the darkly pigmented animal hemisphere and the lighter vegetal region of the amphibian embryo. Forms as a result of cortical rotation triggered at fertilization and indicates the future dorsal side of the embryo.

gut
> Tubular organ lined with endoderm. Functions as the digestive tract of the embryo and the adult. Also gives rise to the respiratory tract, liver, other digestive organs, and the allantois by a process of growing buds or diverticula.

headfold
> The growth of membranes under the head, lifting the head off the yolk to form the subcephalic pocket. Continuous with the lateral body folds.

heart
> Ventral muscular pump of the circulatory system that arises from splanchnic mesoderm.

Hensen's node
> Thickening at the end of the primitive streak, also known as the primitive knot. Homologous to the dorsal lip of the blastopore since the presumptive notochord migrates over it during gastrulation.

hepatic portal system
> System of veins that drain the gut and then go directly to the liver, so that the liver is the first part of the body to see nutrients and contaminants absorbed from the gut.

hindbrain
> The most posterior region of the embryonic brain. Divides into the metencephalon, which gives rise to the cerebellum and pons, and the myelencephalon, which develops into the medulla oblongata.

hindgut
> Portion of the gut caudal to the posterior intestinal portal; gives rise to the cloaca, colon and ilium.

hyaline layer
> Clear gelatinous layer released from cortical granules that surrounds the blastomeres of the sea urchin embryo during cleavage; maintains blastomere adherence.

hypoblast
> Portion of the blastoderm of discoid embryos that lies under the epiblast; formed by the process of delamination. Gives rise to extraembryonic endoderm.

immigration

Process of cell movement through the primitive streak to the interior during gastrulation in the chick embryo.

infundibulum

Gives rise to the posterior lobe of the pituitary. Forms as an evagination from the floor of the diencephalon.

interatrial foramen

Synonym for foramen ovale. Opening in the septum that separates the right from the left atria, allowing blood to pass directly into the left atrium and thus bypassing the pulmonary circulation. Closes soon after birth.

interatrial septum

Wall that separates the right from the left atrium; necessary for division of the circulation into separate pulmonary and systemic systems.

intermediate mesoderm

Synonym for mesomere. A thin bridge of cells connecting the somites and the lateral mesoderm that gives rise to compartments of the urogenital system.

internal gill apparatus

Present in all chordate embryos. Develops into functional gills in fishes, but is modified into other structures in adults of other vertebrates such as the hyoid apparatus, Eustachian tube, thymus, parathyroid and part of the thyroid glands.

interventricular septum

Partition that divides the ventricle into separate left and right compartments. Grows anteriorly from the caudal wall of the ventricle and fuses with the endocardial cushion.

intestine

Portion of gut between the stomach and the anus. Arises from both the foregut and hindgut. Lined with endoderm, the walls are muscular, have an extensive vascular supply, and are highly innervated by the autonomic nervous system.

kidney

Serves as a filter for the blood, removing excess solutes and toxins and excreting them along with water. Three kidney systems may develop: the pronephros, which is the functional kidney of the fish and amphibian embryo; the mesonephros, which is functional in the adult fish and amphibian and the embryonic amniote; and the metanephros, which is the functional kidney of the adult amniote.

laryngotracheal groove

Develops from the floor of the pharynx as a groove that gives rise to the larynx, trachea and lung buds.

lateral body folds

Ventral folding of the somatopleure under the embryo that eventually fuses to form the gut. Delineates the boundary between the embryonic and extraembryonic areas during development.

lateral mesoderm

Mesoderm layer lateral to the intermediate mesoderm. Splits to form the splanchnic and somatic mesoderm, which are separated by the coelom.

lens vesicle

Derived from an ectodermal placode overlying the optic vesicle; differentiates into the lens of the eye.

lens

Spherical body composed of elongated epithelial cells to form an extremely ordered parallel array. Optically clear, it allows light to pass into the eye and focuses the light on the retina.

ligamentum arteriosum

Remnant of the ductus arteriosus that remains after closure of the vessel after birth.

ligamentum teres

Intraabdominal portion of the umbilical vein that remains after its closure at birth, and extends from the umbilicus to the hepatic protal vein.

ligamentum venosum

Remnant of the of the ductus venosus.

liver

Derived from an outpocketing of the gut. During embryogenesis, it is a site of red blood cell development. Largest of the digestive glands, it produces bile, which functions in the digestion of nutrients. The liver manufactures most of the body's proteins, and is a primary site of metabolism and detoxification of many compounds in the body, as well as the site of glucose production.

liver diverticulum

Primordium of the liver, gall bladder and connecting ducts that forms as an envagination of the ventral foregut.

lumbar plexus

Interconnected network of spinal nerves that exit the spinal cord in the region of the hindlimb and innervate the limb.

lung buds

Outpocketings of the laryngotracheal groove that give rise to the primary bronchi.

macrolecithal

Used to describe eggs that are heavily laden with yolk, such as avian eggs.

macromere

The four largest blastomeres of the sixteen-cell embryo.

Malpighian corpuscles

The renal corpuscle, made up of the glomerulus enclosed by Bowman's capsule.

mandibular process

Segment of the first pharyngeal arch that gives rise to the mandible (lower jaw) and tongue.

mantle layer

Intermediate cell layer of the central nervous system. The axons that comprise this layer are devoid of myelin and are therefore gray in appearance.

marginal layer

Outer layer of the central nervous system containing the nerve fibers and neuroblasts. These neurons are heavily myelinated, giving this layer a white appearance.

maxillary process
> The portion of the first pharyngeal arch that forms the upper jaw and palate.

meatus venosus
> Passage through the chick embryonic liver formed by the fused vitelline veins.

mediastinum
> Mass of mesenchyme in which the lungs develop. Persists in the adult as a membranous structure containing the trachea and esophagus that divides the chest cavity into right and left sides.

meiosis
> Process of cell division that occurs during formation of eggs and sperm (gametogenesis); differs from mitosis in that chromosome number is reduced and crossing over of chromosome segments (recombination) occurs.

mesencephalon
> The middle of the three brain vesicles formed in the early embryo. Gives rise to the optic tectum.

mesenchyme blastula
> Stage of sea urchin development characterized by the development and accumulation of primary mesenchyme cells at the vegetal pole of the blastula. The primary mesenchyme cells secrete the embryonic skeleton.

mesenchyme
> Loosely scattered cells that make up the majority of the undifferentiated tissue in the embryo.

mesocardium
> Connective tissue suspending the heart in the body cavity.

mesoderm
> The middle of the three primary germ layers. Gives rise to blood vessels, muscles, connective tissue and kidneys.

mesolecithal
> Eggs with a moderate amount of yolk.

mesomere
> Intermediate-sized blastomeres formed by unequal division during cleavage. Also refers to the intermediate mesoderm.

mesonephric duct
> Duct that connects the mesonephric tubules to the cloaca. First forms as the pronephric duct; retained in male amniotes as the Wolffian duct.

mesonephric tubule
> The functional tubules of the embryonic kidney. Filter blood (via a glomerulus) and produce urine.

metamorphosis
> Radical change in morphology and physiology associated with the transition from a preadult (larval) stage to the adult stage of certain organisms (e.g. frogs, *Drosophila*).

metanephric blastema
The primordium of the third kidney of amniotes. Coalesces around the ureteric bud; gives rise to the kidney tubules.

metanephric kidney
Functional kidney of the adult amniote.

metencephalon
The anterior portion of the rhombencephalon; gives rise to the cerebellum and pons.

micromere
Smallest cells of the cleavage-stage embryo.

midbrain
See mesencephalon.

midgut
Portion of the gut between the anterior and posterior intestinal portals that has not been enclosed by the body folds.

morula
Solid ball of cells that forms after the first several cleavages. Transition to the blastula stage occcurs when the fluid-filled blastocoel cavity forms.

mucous gland
Two glands located ventral to the gills that the frog embryo uses to attach itself to plant material while grazing.

myelencephalon
Most caudal brain vesicle, formed by division of the rhombencephalon. Gives rise to the medulla oblongata.

myocardium
The muscle tissue of the heart.

myotome
Portion of the somite that gives rise to skeletal muscle of the trunk and limb.

nasal pits
Depressions located ventral and anterior to the eye that eventually develop into the sensory epithelium of the nose and the olfactory nerve.

nasal placode
Thickening of the head ectoderm anterior to the eyes. Gives rise to the sensory epithelium of the nose.

nephrostome
Opening of the pronephric tubule into the coelom.

nephrotome
The intermediate mesoderm that gives rise to pronephric, mesonephric and metanephric tubules.

neural crest

Transient, mesenchymal embryonic cell population arising from the dorsal neural epithelium. Cells migrate extensively before differentiating into a variety of structures, including the components of the peripheral nervous system, the adrenal medulla, pigment cells, and much of the craniofacial connective tissue.

neural ectoderm

Portion of the ectoderm that differentiates into neural structures. First apparent morphologically as the neural plate.

neural folds

The edges of the neural plate that rise and fuse together to form the hollow neural tube. Also gives rise to neural crest cells.

neural groove

Indentation between the neural folds.

neural plate

Primordium of the central nervous system. Formed by the thickening and flattening of the dorsal ectoderm.

neural retina

Sensory layer of the eye; develops from the inner layer of the optic cup.

neural tube

Rudiment of the central nervous system formed by fusion of the neural folds. One of the defining characteristics of all chordates.

neurocoel

Lumen of the brain and spinal cord. Develops into the spinal canal and brain ventricles.

neuromere

Segments of the developing brain caused by shallow, annular constrictions of the neural tube; characteristic of the myelencephalon.

neuron

Nerve cell, including the cell body, axons and dendrites.

neurulation

Stage of development during which the neural tube forms.

nose

Area on head (usually the most anterior) containing the olfactory epithelium of the embryo.

notochord

Stiff connective tissue rod of mesodermal origin that lies ventral to the neural tube in chordate embryos.

optic cup

Formed by invagination of the optic vesicle to give rise to a two-layered epithelium. The inner layer gives rise to the sensory retina, while the pigmented retina is derived from the outer layer.

optic stalk

The connection between the diencephalon and the optic cup. Forms a migratory substratum for the neurons derived from the neural retina to follow as they grow toward the brain.

oral plate

Bi-layered epithelium separating the pharynx from the stomodeum. Composed of ectoderm from the stomodeum and endoderm from the pharynx. When the plate opens, the mouth is connected to the pharynx.

otic vesicle

Primordium of the inner ear that forms by invagination of the ectoderm of the otic placode. Found lateral to the myelencephalon.

otocyst

Synonym for otic vesicle.

oviduct

Tube open to the peritoneal cavity that collects and transports the oocyte to the uterus. Site of fertilization in amniotes.

ovum

Synonym for oocyte or egg.

pancreas

Gland of both exocrine and endocrine function found in a fold of the duodenum; arises from the liver diverticulum.

pericardial cavity

Space surrounding the heart, enclosed by the pericardium. Derived from the pleuroperitoneal coelom.

peripheral nervous system

The nervous system excluding the brain and spinal cord. Includes the sensory and motor neurons.

peritoneal cavity

The body cavity enclosing the digestive organs.

perivitelline space

The space between the hyaline layer and the fertilization envelope in sea urchins.

petrosal ganglion

Nerve cell bodies of cranial nerve 9.

pharyngeal arches

Mesenchymal cell masses lying between the pharyngeal clefts. Also known as branchial arches or visceral arches.

pharyngeal cleft

Regions of contact between the pharnygeal pouches and grooves, which perforate to form the gill slits of aquatic species.

pharyngeal groove

Slit-like indentation in the ectoderm which invaginates to contact the evaginations of the pharyngeal pouch. Visible on the surface of the embryo.

pharyngeal pouches

Outpocketings of the endoderm lining the pharynx. In higer vertebrates they give rise to the Eustachian tube (pouch 1), thymus, and parathyroids (pouch 3 and 4).

pharynx
Area of lateral expansion of the anterior (caudal) gut tube that includes the gill apparatus or pharyngeal clefts.

pigmented retina
Differentiates from the outer wall of the optic cup. Non-sensory tissue.

placode
Thickening of tissue, usually ectoderm, that are rudiments of specialized structures including lens, inner ear, and ganglia.

pleural cavity
Space inside the chest enclosed by a membranous sac (the pleura) that contains the lungs. Formed from the pleuroperitoneal cavity. Separated from the peritoneal cavity by the diaphragm in mammals.

pluteus
Free-swimming pelagic larvae of the sea urchin.

polar bodies
Small cells produced by the meiotic division of an oocyte. Contain chromosomes but very little cytoplasm.

polyspermy
Fertilization by more than one sperm. Usually causes abnormal cleavage and death of the embryo.

posterior cardinal vein
Paired veins draining the trunk of the embryo. Empty into the common cardinal veins then into the sinus venosus. Contribute to the renal portal veins in the embryo, before degenerating.

posterior intestinal portal
The opening of the hindgut to the midgut. Moves cranially as the tail fold and lateral body folds fuse, eventually joining the anterior intestinal portal to form the stalk of the yolk sac.

primary mesenchyme
Cells that arise from the vegetal pole of sea urchin blastulae and enter the blastocoel to form the embryonic skeleton.

primitive streak
Cleft in the epiblast of amniotes through which cells immigrate during gastrulation in birds and mammals.

proamnion
Clear area around the head fold of early avian embryos that is devoid of mesoderm.

proctodeum
Inpocket of ectoderm at the posterior end of the embryo that contacts the endoderm to form the cloacal plate.

pronephric duct
Duct to which the pronephric tubules connect. As the pronephric kidney regresses, it serves as the mesonephric duct.

pronephric tubules
Structures in the pronephric kidney through which waste products are channeled to the pronephric duct.

pronephros
The first kidney formed in vertebrate embryos. It serves as the functional kidney of the amphibian embryo, but in higher vertebrates the pronephros forms but never functions, and is replaced by the mesonephros, the functional kidney of the embryo.

prosencephalon
The most anterior of the three brain regions formed in the early embryo that later subdivides into the telencephalon and diencephalon.

pulmonary trunk
Large artery carrying blood from the right ventricle to the lungs. Formed by the longitudinal division of the bulbus arteriosus.

rami
Small branches, for example of a nerve.

Rathke's pouch
Dorsal outpocketing of the stomodeum that meets and fuses with the infundibulum. Forms the anterior lobe, intermediate lobe and pars tuberalis of the pituitary.

rectum
Caudal-most region of the intestine. Absorbs water from the feces.

renal corpuscles
Synonym for Malpighian corpuscles.

rhombencephalon
The hindbrain; subdivides into the metencephalon (anterior) and the myelencephalon (posterior).

rostral
Of or relating to an anterior position.

sclerotome
Ventral portion of the somite that gives rise to the vertebrae and ribs.

segmental plate
Paraxial mesoderm caudal to the last-formed somite. Later subdivides into additional somites.

septum primum
Interatrial septum.

serial sections
Histological sections of an embryo made by slicing the tissue and placing all the sections on slides in anteroposterior order.

sino-atrial valve
Valve between the sinus venosus and the atrium. Prevents back flow of blood to the sinus venosus.

sinus venosus

Posterior portion of early the heart tube. Receives blood from the vitelline veins and common cardinal veins before emptying into the atrium. In the adult it becomes part of the right atrium and gives rise to the sino-atrial node, which regulates the heartbeat.

somatic mesoderm

Mesodermal cells that arise from the lateral plate mesoderm and contribute to formation of the body wall. Also contributes to the extraembryonic chorion and amnion.

somatopleure

Combined layers of ectoderm and somatic mesoderm. Forms the amnion and chorion and contributes to the body wall.

somites

Blocks of mesodermal tissue arranged on both sides of the spinal cord. Contributes to the vertebrae, the axial musculature, and the dermis.

sperm

Fully-differentiated male gametes.

spinal cord

Portion of the hollow nerve tube caudal to the brain. Part of the central nervous system.

spinal nerve

Nerves exiting the spinal cord that contain preganglionic axons of the sympathetic nervous system and sensory and motor fibers of the somatic nervous system.

splanchnic mesoderm

Mesodermal cell layer that associates with the endoderm. Contributes to the wall of the gut, the respiratory tract, the heart, the extraembryonic yolk sac, and the allantois.

splanchnopleure

Combined layers of endoderm and splanchnic mesoderm. Forms the yolk sac, allantois and gut wall.

stomach

Enlarged muscular sac between the esophagus and intestine. Functions in digestion of food, especially proteins.

stomodeum

Primordium of the mouth, formed by invagination of the head ectoderm. It contacts the anterior end of the endodermal tube to form the oral plate.

subcardinal anastomosis

Connection between the right and left subcardinal veins. Contributes to the posterior vena cava.

sub-cephalic pocket

Space beneath the head of the embryo, formed by the head fold of the amnion.

subgerminal cavity

Space between the hypoblast and yolk mass in early avian embryos.

tail fold

Ventral folding of the ectoderm, mesoderm, and endoderm under the posterior end of a discoid embryo to form the tail and hindgut.

telencephalon
Most anterior of the brain vesicles. Gives rise to the cerebrum and the olfactory bulbs.

teratogen
Compound that, when absorbed by the body of a pregnant female, can cause abnormal development of the fetus.

teratology
Study of abnormal development.

thyroid gland
Endocrine gland that arises as a diverticulum of the ventral pharynx at the level of the 2nd pharyngeal arch. Produces thyroxine, which regulates the basal metabolic rate of the body.

trachea
Tube connecting the pharynx to the bronchi. Arises as a diverticulum of the pharynx.

transverse septum
Separates the pericardial cavity from the peritoneal cavity.

trigeminal nerve
The 5th cranial nerve. Derived from neural crest cells and neural ectoderm. Supplies the sensory innervation of the face.

tubular heart
The early heart prior to its separation into paired chambers.

tympanic cavity
The middle ear; derived from the first pharyngeal pouch.

umbilical vein
Embryonic vessel carrying oxygenated blood to the embryo from the placenta or from the allantois.

ureteric bud
Outpocket of the mesonephric duct. Gives rise to the renal pelvis, collecting ducts, and ureters. Induces development of the metanephric blastema.

Urochordata
Subphylum of the Chordates that includes the tunicates.

uterine horn
Paired elongated tubular part of the uterus. Site of implantation and development of the fetus in most mammals.

vagus nerve
Cranial nerve 10. Mediates much of the parasympathetic autonomic input to the digestive tract.

vegetal pole
The yolk-laden pole of the egg opposite the animal pole in anamniotes.

ventral ramus
Ventral branch of spinal nerve. Innervates the body wall and viscera, as well as the limbs.

ventral aorta
> Drains the heart in the early embryo. Connects the bulbus arteriosus with the aortic arches. Later develops into the ascending aorta.

ventral root
> The portion of the spinal nerve carrying motor and preganglionic sympathetic fibers to the periphery.

ventricle
> 1. Largest chamber of the embryonic heart. Eventually becomes partitioned into left and right chambers, which pump blood through the pulmonary or systemic circulation.
>
> 2. Enlarged regions of the brain's lumen containing cerebrospinal fluid.

Vertebrata
> Subphylum of the Phylum Chordata that includes all vertebrates: mammals, birds, fishes and amphibians.

visceral arch
> Synonym for pharyngeal arch. The mesenchymal tissue primarily of neural crest origin between the pharyngeal pouches.

vitelline arteries
> The arteries supplying blood to the yolk sac. Derived from branches of the dorsal aorta. Give rise to the mesenteric arteries.

vitelline envelope
> A thin glycoprotein layer surrounding the egg. In the sea urchin it lifts off the plasma membrane to form the fertilization envelope.

vitelline veins
> Veins draining the yolk sac. Fuse to form the sinus venosus. Also form the meatus venosus as they pass through the liver. Also give rise to the hepatic portal vein.

Wolffian duct
> Synonym for the mesonephric duct. Gives rise to the tubular portions of the male reproductive tract: the epididymis, the vas deferens, and the ureter.

yolk endoderm
> Large, yolk-filled cells derived from the vegetal hemisphere that form the gut.

yolk plug
> Yolk-laden cells that protrude from the blastopore during gastrulation in the amphibian embryo. Forms the yolk endoderm.

yolk sac
> Splanchnopleure-derived extraembryonic membrane surrounding the yolk in avian embryos. Yolk sac cells absorb nutrients from the yolk and transport them to the embryo via the vitelline veins. Also serves as a respiratory organ in the early chick embryo.

zygote
> Fertilized egg.